Res cogitans extensa

European University Studies
Europäische Hochschulschriften
Publications Universitaires Européennes

Series XX
Philosophy

Reihe XX Série XX
Philosophie
Philosophie

Vol./Bd. 744

PETER LANG
Frankfurt am Main · Berlin · Bern · Bruxelles · New York · Oxford · Wien

Georg Theiner

Res cogitans extensa

A Philosophical Defense of the Extended Mind Thesis

PETER LANG
Internationaler Verlag der Wissenschaften

Bibliographic Information published by the Deutsche Nationalbibliothek

The Deutsche Nationalbibliothek lists this publication in the Deutsche Nationalbibliografie; detailed bibliographic data is available in the internet at http://dnb.d-nb.de.

Printed with financial support of the Federal Ministry of Science and Research in Vienna.

ISSN 0721-3417
ISBN 978-3-631-57937-4

www.peterlang.de

For Allison,
with love

Just as you cannot do very much carpentry with your bare hands, there is not much thinking you can do with your bare brain.

–Bo Dahlboom, "Computer Future"

Before the day is done, however, I hope to convince you at least of this: that the old puzzle, the mind-body-problem, really involves a hidden third party. It is the mind-body-*scaffolding* problem. It is the problem of understanding how human thought and reason is born out of looping interactions between material brains, material bodies, and complex cultural and technological environments. We create these supportive environments, but they create us too. We exist, as the thinking things we are, only thanks to a baffling dance of brains, bodies, and cultural and technological scaffolding.

–Andy Clark, *Natural Born Cyborgs*

Thinking is a kind of building, a kind of intellectual niche construction that appropriates and integrates material resources around one into pre-existing cognitive structures. In cognition, agents modify or augment the capacities that those pre-existing structures enable.

–Rob Wilson & Andy Clark, "How to Situate Cognition"

Spirit is real only when by its own motion it divides itself, gives itself limit and finitude in the natural needs and the region of external necessity, and then, by moulding and shaping itself in them, overcomes them, and secures for itself an objective embodiment. […] Its externality thus in the first instance receives the rationality, of which it is capable, namely, the form of universality characteristic of the understanding. Only in this way is spirit at home and with itself in this externality as such. […] Culture or education (*Bildung*) is, as we may thus conclude, in its ultimate sense a liberation, and that of a high kind. […] In the individual agent this liberation involves a struggle against mere subjectivity, immediate desire, subjective vanity, and capricious liking. The hardness of the task is in part the cause of the disfavour under which it falls. None the less is it through the labour of education that the subjective will itself wins possession of the objectivity, in which alone it is able and worthy to be the embodiment of the idea.

–Georg Wilhelm Friedrich Hegel, *Philosophy of Right*, §187

Preface

About ten years ago, I attended one of Andy Clark's talks about the "extended mind" thesis. The thesis asserts that a significant portion of human cognition literally extends beyond the brain into the body and the world, actively incorporating environmental props, tools, and scaffolds. I had barely started my Ph.D. in Philosophy and Cognitive Science, and was still under the impression that Noam Chomsky, Jerry Fodor, and Ned Block had essentially figured out the human mind. In other words, I was in many ways committed to some broad form of "Cartesian materialism." Needless to say, I was not fully convinced by Clark's brand of "active externalism," even though the content of his talk clearly resonated with me. In particular, I was intrigued by its unexpected anti-Cartesian twist. It seemed to offer an interesting alternative to the traditional choice between Cartesian dualism and Cartesian materialism. Perhaps the mind is neither a *res cogitans* that is fundamentally distinct from any physical *res extensa*, nor is it identical with the brain. What if the reason why human cognition sticks out in the animal world like an elephant's trunk is its remarkable ability to extend itself into the world, i.e., to become a *res cogitans extensa*? This powerful idea struck me as a potentially game-changing move in cognitive science, because it promised to reconcile a deep neural continuity with an equally profound cognitive discontinuity between human beings and their closest biological kin. The discreet Hegelian charm of the "extended mind" motive enticed me to undertake a systematic study of its empirical credentials and philosophical underpinnings. This book is the first part of that larger project. Its goal is to clarify contemporary versions of the extended mind thesis, and to defend its central insight that cognition is not confined to the boundaries of the biological individual.

This book has been written over the course of several years, during which my views about situated, embodied and extended cognition have been shaped by many people. I would like to extend special thanks to Rob Wilson, who was the greatest post-doctoral advisor I could have asked for, and to Andy Clark, for kindling my interest in the extended mind thesis. At Indiana University Bloomington, I was blessed with an intellectually diverse and exceptionally helpful doctoral committee that consisted of Timothy O'Connor, Robert Goldstone, Fred Schmitt, and Jonathan Weinberg. I have also benefited from stimulating conversations about the extended mind with Darren Abramson, Kenneth Aizawa, Colin Allen, Olle Blomberg, Jordi Cat, Tony Chemero, Paul Eisenberg, Steve Harris, David Landy, Jun Luo, Lars Marstaller, Sergio Martínez, Adam Morton, David Olson, Robert Rupert, Dan Ryder, Brian Smith, Michael Spivey, John Sutton, Manuela Ungureanu, Bill Wimsatt, and Carlos Zednik (apologies to anyone I have forgotten!). For financial support, I am very grateful for the Izaak W. Killam postdoctoral fellowship that I held during the two years of my stay at the University of Alberta (2008-2010).

Finally, I would like to thank my immediate family for supporting and also fostering my philosophical pursuits. I am especially grateful to my wife, Allison, not only for proofreading the entire book, but for keeping my life in proper perspective and balance. My parents, Liselotte and Klaus, have been an invaluable source of support throughout my entire education and always came through for me when I needed them. I would also like to give my warmest thanks to my parents in-law, Doug and Peggy, for their hospitality and encouragement. Last, but not least, let me give epistemic credit to my multi-scaffolded environment, especially my dependable and always accessible notebook, which gave me the extra shot of cognitive fuel that allowed me to write this book.

Georg Theiner
Vienna, April 2011

Table of Contents

1. Extending the mind

1.1 Externalism: mortification or empowerment?

There is a wide-spread picture of the mental as something *inner* – a view whose influence on our thinking about the mind is as tenacious as it is habitually taken for granted. Having a mind, according to this picture, is a property of individual subjects, in the sense that what's going on in somebody's mind (at a given time) is settled entirely by intrinsic properties of that individual. It's entirely up to *me* what's going on in *my* mind. I do not have to look *outside* myself in order to settle the question of what I am currently thinking. More generally speaking, the nature of a subject's mental state is logically independent of the subject's relationship to its environment, including other subjects. The acceptance of this picture, which is known as *psychological internalism* in contemporary analytic philosophy of mind, runs much deeper than what could be captured by any explicit statement or doctrine. But like many other influential ideas, it is through the philosophical writings of René Descartes that psychological internalism has left its distinctive mark on our intellectual landscape – enough so to warrant the label "Cartesian" internalism. Cartesian internalism is alive and kicking in our best scientific attempts to understand the mind even though Descartes' mind-body dualism – the other backbone of the Cartesian legacy – has (at least officially) long been expunged. Contrary to what Descartes believed, the bulk of mainstream research in psychology, cognitive science, and cognitive neuroscience is predicated on the materialist assumption that the mind is – albeit a very special – a kind of physical thing, open in principle to the same kind of objective investigation as any other natural phenomena such as spiral nebulae, photosynthesis, and tuberculosis. Yet under the scepter of materialism, Cartesian internalism remains the reigning view in this enlightened day and age of scientific psychology. How did Cartesian internalism so splendidly survive its materialist transformation?

As a substance dualist, Descartes argued in the *Meditations* that minds and bodies have fundamentally different natures (Descartes, 1641/1996). A body and the organs from which it is composed – such as its heart, liver, and brain – is a physical substance whose essence it is to take up space – its *extension*. A physical substance operates purely on the basis of mechanical principles. A mind, on the other hand, is essentially a thinking thing, governed only by the laws of reason. For Descartes, this meant that minds must be an altogether different, non-physical kind of substance which lacks spatial extension. Descartes' strong form of substance dualism is incompatible with a fundamentally materialistic world view (Kim, 1996). The main thrust of Cartesian internalism, on the other hand, is largely independent of its dualist cousin. Following Rowlands (2003, pp. 12-18), we can define Cartesian internalism as the combination of two logically distinct theses, even though the two are frequently held together. The first thesis concerns the *location* of mental phenomena;[1] the second concerns the conditions under which a subject can be truly said to *possess* certain mental phenomena:

LOCATION CLAIM: Any mental phenomenon *M* is spatially located entirely inside the boundaries of the subject *S* of *M*.

POSSESSION CLAIM: The possession of any mental phenomenon *M* by a subject *S* does not logically depend on any feature that is external to the boundaries of *S*.

We can bring out the difference in truth conditions between these two claims with a mundane example – the property of having sunburn (Davidson, 1987). Sunburn is a burn of skin tissue which is produced by an overexposure to ultraviolet radiation – most commonly caused by the sun's rays, or perhaps due to artificial sources such as tanning lamps. Whether some condition of the skin is correctly identified ("individuated") as sunburn or not arguably depends on its causal etiology – only the overexposure to incident UV-radiation can cause sunburns. For instance, let us suppose that there is a genetic disease which somehow triggers the same chemical reaction of the skin as a result of (say) taking a foam bath. Such a condition would not equally count as an instance of sunburn. Having sunburn – its possession conditions, as we say – is thus crucially dependent on "distal" causal-historical facts concerning events that occur outside the boundaries of the skin. However, even if the possession of sunburn is individuated in part by an organism's relation to its environment, it evidently does not follow that the token piece of sunburnt tissue itself is located outside the skin of the organism. Using similar examples, we can easily see that the location-claim and the possession-claim of Cartesian internalism do not logically entail each other (Hurley, 1998a; Rowlands, 2003).

Let us now aim for a more succinct characterization of Cartesian internalism that meshes with current materialist thinking about the mind. In a way, this makes our task easier, as it is not altogether obvious how the above formulation of the location-claim can be fully squared with Descartes' own mind-body dualism. In particular, if minds are non-physical substances, how could they have a spatial location at all? Yet having a location would certainly seem to be required for us to draw a distinction between what's "inside" and "outside" the bounda-

1 I use the expression phenomenon here as a generic term to denote all kinds of things mental—including mental particulars, states, properties, processes, events, dispositions, etc.

ries of a subject. One way of addressing this problem is to understand Cartesian minds as metaphysical equivalents of geometrical points (Campbell, 1984, pp. 45-48). Points do indeed have a spatial location, even though they do not occupy any space.[2] This might allow us to say that Cartesian minds, and thus the manifestation of their states, literally have a location somewhere inside the pineal gland (i.e. the part of the brain which Descartes believed to be the locus of causal interaction between body and mind). Even so, it would remain unintelligible to say that mental states are spatially located "inside" the boundaries of its *bearer* if that bearer lacks spatial extension.

A much more straightforward, contemporary interpretation of Cartesian internalism appeals to the location of mental states with respect to the boundaries of a *person*. Whereas Descartes famously struggled to give adequate expression to his notion of a person as a "union" of mind and body, present-day materialists have no qualms about identifying the person with the physical, organism-bound individual. If we consider the individual organism as the proprietary subject of mental states, it seems (at least prima facie) perfectly intelligible to assert that all mental phenomena happen *inside* the boundaries of the subject – whether those are taken to be the boundaries of the brain, the entire central nervous system, or the organism as a whole. In short, as Harnad (2005, p. 502) has recently put it – and as most materialistically-minded scientists and philosophers would agree – the hypothesis that a person's mind is inside her head is a bet "no riskier than the bet that we have heads at all." I take this to be the most intuitive rendering of the internalist location-claim.

Let us briefly turn to the internalist possession-claim, because it is also related to Descartes' arguments for substance dualism. In the *Meditations*, Descartes argues that the mind is only contingently related to physical things on the grounds that changing the properties of anything physical leaves the essential properties of the mind untouched. This means that the possession of a mental state must be exhaustively determined by the intrinsic nature of the mind, and cannot metaphysically depend on its relations to anything physical. For the present-day Cartesian materialist, a very similar line of reasoning purportedly shows that nothing outside a person's body makes a constitutive (i.e., non-causal) difference to what mental states it undergoes (at a given time). To put things in the familiar dramatic idiom: what the internalist materialist claims is that it is metaphysically possible for a disembodied "brain in a vat" (Putnam, 1981) to duplicate the exact same mental properties that are had by a normal, embodied subject abiding in a normal environment, notwithstanding the enorm-

2 This would still leave unsolved the so-called "monogamy"-problem for Cartesian souls; see Kim (2001).

ous differences with respect to the causal and/or historical relations between the envatted and the embodied-cum-embedded brain. According to the possession-internalist, the conceivability of the brain-in-a-vat scenario shows that extrinsic facts concerning the causal and/or historical relations between the organism-bound subject and its environment cannot be metaphysically constitutive features of mental states. In sum, the gist of the internalist possession-claim is that mental properties are not *individuation-dependent* on environmental states of affairs (Rowlands, 2003; McGinn, 1989).

In the past few decades, possession-internalism has come under attack by a cluster of philosophical theories frequently grouped together under the banner of *content-externalism*. According to this view, mental states are partly individuation-dependent on environmental factors, because the *content* of those mental states – what they are about – is crucially determined by external relations which a subject bears vis-à-vis those factors. Support for this view has originally been derived from a series of thought experiments first proposed in the 1970s (Putnam, 1975b; Burge, 1979; Burge, 1982). Challenging traditional "Cartesian" intuitions, content-externalists claim that if we hold fixed everything that is going on inside a subject's head, we can still change what kind of mental states the subject undergoes just by varying certain aspects of her physical (Putnam) or social (Burge) environment. This difference in mental state type, so content-externalists would argue, can occur in the absence of any intrinsic changes to the organism-bound subject. If the meaning of a mental state "ain't in the head" (Putnam, 1975b), we thus crucially have to take into account those beyond-the-head factors to determine what a person is thinking. The arguments for content-externalism, if cogent, have significant implications not only for the nature of mental states, but for linguistic meaning and reference, our knowledge of the external world and the coherence of philosophical skepticism, the nature of self-knowledge, and the problem of mental causation (for overviews, see Lau & Deutsch, 2010; Kallestrup, 2010; Rowlands, 2003).

However, there are also significant limitations to the scope of content-externalism. First, content-externalism only concerns the possession-claim of Cartesian internalism. Even if content-externalists are right that the representational *content* of mental states (in particular, their truth-conditions) is at least partly determined by a subject's relation to the environment, this would most plausibly only affect the possession of representational mental states that are semantically evaluable in a roughly propositional sense, such as beliefs and desires (the so-called "propositional attitudes"). In contrast, experiencing a certain sensation such as a tickle, an itch, or an orgasm has a distinctive feel that does not seem to be intrinsically directed at particular states of affairs, unlike (say) the belief that Henry VIII was an English king. Conversely, there is no distinctive way it *feels* to have the belief that Henry VIII was an English king. Hence if the purely qualitative feel of phenomenally conscious mental states is not exhausted by their representational content, the "what-is-it-like-ness" (Nagel, 1974) of those mental states would lie outside the scope of content-externalism.

There are at least two further reasons why traditional arguments for content-externalism, which were largely driven by the goals and demands of conceptual analysis, have not had much of an impact on the explanatory practices of cognitive science (Hurley, 1998a; Rowlands, 2003). For one, even representational mental states would be individuation-dependent on a subject's environment only if it can be shown that semantic properties are part of their essence as mental states (of a certain type). But that assumption is not universally accepted. In cognitive science, the notion of a *mental representation* is usually treated as a technical term, especially within the Computational Theory of Mind (Horst, 2011). It refers to the occurrence, storage, and transformation of information-bearing structures in the mind/brain that are causally implicated in cognitive processes such as categorization, language, and reasoning (Thagard, 2005). While mental representations in this sense contribute to the formation of propositional attitudes (e.g., the fixation of beliefs), they are typically not consciously accessible, nor do they have to be semantically evaluable in the same way as propositional attitudes (Stich, 1978). Hence there has been considerable debate whether mental representations in cognitive science are (or at least ought to be) individuated externally or purely "solipsistically" (Fodor, 1980), i.e., in a way that does not constitutively depend on the existence of certain entities in the environment.[3] Finally, as the sunburn example nicely shows, even assuming that the representational *content* of an intentional mental state is partly dependent on environmental states of affairs, this does not imply that the representational *vehicle* of that state is located outside the head. Since it is clearly an important goal of cognitive science to explain the physically constitutive vehicles of cognition, one can see how the philosophical revolt of content-externalism has at times appeared to psychologists more like a tempest in a teapot.

My goal in this book is to explore a different, and potentially far more momentous type of externalism which is known under various guises such as the *extended mind thesis* (Clark & Chalmers, 1998; Clark, 1997, 2003, 2008), *locational externalism* (Wilson, 1995, 2004), *active externalism* (Hurley, 1998a, 1998b), *enactivism* (Varela, Thompson, & Rosch, 1991; Noë, 2004, 2009; Thompson, 2007), *cognitive integrationism* (Menary, 2006, 2007; Sutton, 2006, 2010), or *environmentalism* (Rowlands, 1999). In his new flagship presentation,

---

3 Classical discussions of content-externalism can be found in Burge (1986), Segal (1989, 1991), Egan (1992), and Shapiro (1997).

Supersizing the Mind, Andy Clark (2008) has put the extended mind thesis quite succinctly as the claim that thinking doesn't happen only in our heads but that "certain forms of human cognizing include inextricable tangles of feedback, feed-forward and feed-around loops: loops that promiscuously criss-cross the boundaries of brain, body and world" (p. xxvi). The underlying theme of this "active" form of externalism is that thinking is a kind of intellectual niche construction – activities in which cognitive agents create, appropriate, and integrate material and social-cultural resources of their environment into their pre-existing cognitive structures, and thereby re-configure the capacities which those structures enable (cf. Wilson & Clark, 2009, p. 58).

For example, human beings are uniquely talented in modifying their environments through the creation of cognitive artifacts (Norman, 1991) – tools which make us smarter by transforming the nature of otherwise difficult cognitive tasks into something more tractable for the biologically basic capacities of our mammalian brains. Some of those artifacts give us access to hitherto unavailable kinds of information (e.g., compass, sextant). Others help us to offload, re-represent, and return information (e.g., pen and paper, language, gesturing, writing systems), and yet others function as active sources of external information-processing and representation-transformation (e.g., abacus, pocket calculator, or word processor). In sum, cognitive artifacts enable us to distribute cognitive activities across time, space, and people (Hutchins, 1995a; Hollan, Hutchins, & Kirsh, 2000; Wilson & Clark, 2009). They make us smarter by transforming the nature of otherwise difficult cognitive tasks into something more tractable for the biologically basic capacities of our mammalian brains. The skillful incorporation of environmental resources into our problem-solving routines is often accompanied by corresponding changes in our neural machinery (Clark, 2003, 2008). As we create and immerse ourselves in artifact-enriched environments, these neural changes reflect our tacit, automatic reliance on their continuous bio-external availability. If the cognitively relevant interactions between brain, body, and environmental resources are sufficiently intimate and functionally integrated in the context of performing a cognitive task, it becomes increasingly arbitrary to single out the contributions of the body or the external world as mere "inputs" or "instruments" for cognition. As a case in point, when historian Charles Weiner found pages of Nobel Prize-winning physicist Richard Feynman's notes, he remarked that they were a fine "record" of Feynman's day-to-day work. Feynman himself, however, insisted that the notes were not a record of his work but where he actually *did* the work (reported in Clark, 2008, p. xxiii). Clark (loc. cit.) suggests that Feynman was literally *thinking* on paper, in the sense that the loop through pen and paper was part of the physical machinery responsible for shaping the flow of Feynman's thought processes. According to the extended mind thesis, cognition itself extends beyond the head into the body and the external world.

Unlike the somewhat more conservative brand of content-externalism, the extended mind thesis directly targets the location-claim of Cartesian internalism. Its cultural roots lie mostly outside academic philosophy. Important precursors of its underlying emphasis on cognition as a concrete material, embodied, and situated activity can be found – among many others – are the works of the psychologists Lev Vygotsky and James J. Gibson, the biologists Jakob von Uexküll and Francisco Varela, and the cognitive scientists Edwin Hutchins, Donald Norman, and David Kirsh. But of course there are many noteworthy philosophical antecedents of active externalism, united in their otherwise divergent approaches – as one might expect – by the pursuit of distinctively anti-Cartesian agendas. Various patterns of thought which fit into the larger themes of the extended mind thesis have been explored, in one way or another, by such writers as Hegel, Marx, Nietzsche, Heidegger, Merleau-Ponty, Dewey, James, Wittgenstein, and Dennett.[4]

What are the larger philosophical implications of the extended mind thesis? Sigmund Freud (1955) famously described as "mortifications of humanity" three paradigm-shifting discoveries of modern science which, or so he presumes, must have deeply offended our collective narcissism, which is why they all have been met with great resistance to the point of utmost hostility.[5] The first mortification refers to the de-centering of man's place in the universe, a cosmological blow dealt to us through the Copernican revolution. The second mortification refers to the biological blow which results from the Darwinian insight that man was not specially created by God, in his image, but is of common descent with other life forms on our planet. The third mortification refers to the psychological blow which occurs as the ego discovers – sparked, of course, by Freud's own psychoanalytic theory – that it is not even master "in its own house," but constantly at risk of being invaded by uncontrollable libidinal drives and the alien forces of the unconscious.

Within analytic philosophy, content-externalism has at times been met with the obstinate resistance of a mortified self which refuses to believe that the mind

---

4 For historical overviews of some important scientific and philosophical antecedents of the "situated cognition" movement, see Clancey (1997, 2009), Wheeler (2005), Rockwell (2005), Chemero (2009), and Gallagher (2009).

5 I am using Freud's claims merely as an expository device. Apart from the obvious fact that Freud was trying to boost his own image as a "mortifier of humanity," there are other reasons to doubt the historical accuracy of Freud's observations about how these discoveries were actually received; see Pauen (2007).

is not the master of the content of its own thoughts. It is a common Cartesian sentiment that content-externalism so flagrantly encroaches upon our impervious sense of first-person authority that it ought to be rejected. The new brand of active externalism is likely to engage the same unconscious defense mechanisms of the Cartesian materialist, perhaps even more fervently so. Just as it was evidently true for Descartes that the mind is an immaterial substance, it seems so evidently true to the Cartesian materialist that the mind must be located inside the brain. It seems as if the brain – perhaps the final frontier of our scientific age – has become the new refuge to which our narcissistic egos have had to retreat. But the extended mind thesis claims that the physically constitutive mechanisms of human-sized cognition leak beyond the brain into the body and its material, social, and cultural environment. If this is true, brains would finally lose their exclusive rights to be featured as the only constitutive producers of mental phenomena.

Is the extended mind thesis the latest mortification of humanity, or does the extrusion of the mind into the world reveal the ultimate source of human cognitive empowerment? Read on and see for yourself.

1.2 How to classify the DEEDS of cognition

Since its beginnings, the emerging field of cognitive science has been periodically hit by waves of criticisms issued by philosophers, psychologists, sociologists, anthropologists, and cognitive scientists alike who are keen on showing that cognitive science – as currently practiced – is somehow fundamentally flawed. A recently growing concern has been that mind and cognition ought to be understood as dynamic, embodied, embedded, distributed, and situated ('DEEDS[6]') phenomena. While paradigms shifts are certainly nothing unusual for the tectonic of a developing science (Kuhn, 1970), the penchant for revolutions seems to be particularly notable feature of the comparatively young field of cognitive science – a field whose official inception (at least in its contemporary form) only dates back a few decades to the so-called "cognitive revolution" of the mid-1950s. This has two related consequences. More so than in other, more established sciences, the main gesture of its revolutionary movements is often to say primarily what cognition is *not*. This has undeniably cathartic effects, and also helps to bring into sharper focus the reigning orthodox wisdom about the mind. But it can also make it hard to pin down the substantive content of the positive view which is proffered as an alternative theory of the mind. As we shall see in this section, it is not easy to disentangle the different threads that are frequently interwoven in the DEEDS-literature. These threads span over a large

6 This acronym was coined by Marsh and Onof (2008).

and heterogeneous array of research frameworks and methodologies which often do not bear more than family resemblances to one another (Robbins & Aydede, 2009).

My goal in this chapter is to review several recurring themes within the DEEDS-agenda that will help to bring out the basic contours of the extended mind thesis. In my outline, I shall try to distinguish – as much as this is both conceptually feasible and practically fruitful – between distinctively "situated/embedded," "embodied," and "distributed" aspects of cognition, and corresponding approaches to cognitive science.

1.2.1 "Situated/embedded" approaches to cognition

In his respective entry to the *MIT Encyclopedia of Cognitive Science*, Brian Smith characterizes the "situated/embedded" movement with a relative shift in emphasis on what are considered to be the defining features of human cognition:

> "Situatedness arose in the 1980s as a reaction against the then-dominant classical view of mind. The classical approach, inherited from the logical and metamathematical traditions (dubbed "GOFAI" by Haugeland (1997), for "good old fashioned artificial intelligence"), views cognition as: *individual,* in the sense that the essential locus of intelligence is taken to be the solitary person; *rational,* in that deliberative, conceptual thought is viewed as the primary exemplar of cognition; *abstract,* in the sense that implementation and the nature of the physical environment are treated as of secondary importance (if relevant at all); *detached,* in the sense that thinking is treated separately from perception and action; and *general,* in the sense that cognitive science is taken to be a search for universal principles of general intellection, true of all individuals and applicable in all circumstances.
>
> Situated approaches reject one or more of these assumptions, arguing instead that cognition (indeed all human activity) is: *social,* in the sense of being located in humanly constructed settings among human communities; *embodied,* in that material aspects of agents' bodies are taken to be both pragmatically and theoretically significant; *concrete,* in the sense that physical constraints of realization and circumstance are viewed as of the utmost importance; *located,* implying that context-dependence is a central and enabling feature of all human endeavor; *engaged,* in that ongoing interaction with the surrounding environment is recognized as primary; and *specific,* in that what people do is seen as varying, dramatically, depending on contingent facts about their particular circumstances" (Smith, 1999, p. 769).

As we can see, Smith treats the situatedness/embeddedness of cognition as a gradient of how "deeply" or "constitutively" we take cognitive processes to depend on the environmental context in which they occur. He points out that the theoretical emphasis on the context-dependence of cognition has several important implications for our cognitive-scientific practice. First, adopting a situated approach reorients our decision to consider certain cognitive activities as paradigmatic exemplars of human intelligence. For instance, instead of the traditional emphasis put on the abstract, "ratiocinative" provinces of cognition, the situationist focuses on the cognitive achievements that underlie everyday activities such as cooking dinner, making one's way across a crowded subway platform, or negotiating turn-taking in a conversation. Second, this shift is underpinned by a rejection of the "Cartesian" assumption that the central domains of human thinking are relatively autonomous "inner" capacities that can be studied in isolation from our sensory and motor interactions with the physical world. Third, focusing on the dynamics of cognition *in situ* requires a concomitant shift from "in-vitro" to "in-vivo" methods insofar as psychologists will have to leave behind the artificially constrained environments of their labs if they want to study the behavior of people acting in real-world situations. Fourth, the methodological implications of the situated turn are reflected in a change of disciplinary affiliations. A situated/embedded-reformed cognitive science is prone to reject the traditional dominance of logic, computer science, and lab-based psychology, and become much more open to research in sociology, anthropology, ethology, science studies and related fields – in a joint attempt to create what Clifford Geertz (1983) once nicknamed an "outdoor psychology."

1.2.2 "Embodied" approaches to cognition

Margaret Wilson (2002) uses a different, "divide-and-conquer" approach to pick apart the flurry of ideas that are frequently run together under the banner of "embodied cognition," as if they characterized a single, monolithic point of view.[7] In her critical review of the field, Wilson distinguishes between six importantly different claims.

1. Cognition is situated. One meaning of "situated cognition" refers to the interactive type of "on-line" cognition that is directly geared to the accomplishment of specific cognitive tasks in concrete settings. Situated cognition in this sense is continually driven by incoming perceptual information, and directly results in motor activity that affects the environment in which it takes place (Clark, 1997; Thelen & Smith, 1994; Chiel & Beer, 1997; Pfeifer & Scheier, 1999; Pfeifer, Bongard, & Grand, 2007). Wilson's examples here include driving during rush hour, holding a conversation, and moving around a room in the course of planning how to rearrange one's furniture (Wilson, 2002, p. 626). According to this characterization, any "off-line" cognitive activities that are not concerned with processing immediate input from one's local environment (e.g., planning the next summer vacation) would thus not count as situated.

7 For a somewhat more ecumenical survey, see Anderson (2003).

2. Cognition is time-pressured. The claim that cognition is situation-bound is sometimes taken to imply that agents must perform their tasks under severe time constraints (Port & van Gelder, 1995; Pfeifer & Scheier, 1999; Brooks, 1999). The necessity of coping with imminent time-pressure has been downplayed (if not neglected) by traditional "rationalist" conceptions of intelligence, where it is often assumed that the nature of our core cognitive competencies is best understood in abstraction from performance factors such as time constraints (Franks, 1995). Theories of cognition which are based on such idealizations, so the situationists argue, do not adequately describe the cognitive architecture of creatures like us who have primarily evolved to act and react smoothly in real time. Cognitive domains in which the continuous reciprocal interaction between agents and their environments is clearly essential include all kinds of tasks which heavily rely on perceptuomotor coordination, such as the planning and execution of skilled hand and finger movements, playing tennis, or changing lanes while driving in heavy traffic (Wilson, 2002, p. 628).

3. We off-load cognitive work onto the environment. When we are asked to perform cognitively taxing tasks, we are quickly reminded of our biologically imposed cognitive limitations. For example, in unfamiliar situations that demand fast and constantly changing behavioral responses, our cognitive system faces a severe "representational bottleneck" (ibid., p. 628). There is simply not enough time to represent all the potentially relevant aspects of our environment in a way that would allow us to choose the most appropriate plan of action. To deal with this cognitive limitation, an embodied strategy is to minimize the perceptual intake of information on a need-to-know basis, using the world as "its own best model" (Brooks, 1997, p. 395; see also Ballard et al., 1997; O'Regan & Noë, 2001).

A striking demonstration of cognitive off-loading comes from an experimental investigation of how players of the video game Tetris make time-pressured spatial decisions about where they should move a falling block (Kirsh & Maglio, 1994). It is of critical importance that the decision of whether a falling block fits into one of the sockets at the bottom of the screen is made fast, before the emerging structure impedes further operations that would be necessary to move the block to the desired location. Kirsh and Maglio found that the improved performance of expert players on this task results from directly rotating the falling blocks on the screen, rather than rotating mental images of those blocks in their heads, which is what novices do. For the expert player, the action of manipulating the falling tiles on the screen thus performs multiple functions (Kirsh & Maglio, 1994). On the one hand, it serves a *pragmatic* purpose, because the manipulations help her to achieve the physical goal of moving a tile into one of the existing sockets. But in addition, her manipulations also serve an *epistemic* purpose, because they transform a cognitively taxing spatial reasoning task into a much simpler visual pattern-matching task.[8]

Wilson (2002, p. 629) also draws an important distinction between *situated* and *symbolic* varieties of cognitive off-loading. In Kirsh & Maglio's example, one and the same action does pragmatic and epistemic double-duty, because the desired physical goal state and the epistemic goal state of a Tetris player are essentially the same. A tile that fits into one of the sockets not only helps to build a larger, uninterrupted structure (which increases one's score), but also solves the spatial reasoning problem of how to build such a structure. The sameness of pragmatic and epistemic goal states makes the Tetris-case an instance of *situated* off-loading. Other familiar examples of this sort arise when we exploit the spatial structure of our environment to solve a variety of cognitive tasks that are themselves spatial in nature. For instance, we walk around a room to generate and evaluate plans for how to rearrange our furniture; or, when we give someone directions for how to get somewhere, we first turn ourselves and our listener into the appropriate direction; similarly, we lay out the pieces of a construction that needs to be assembled in roughly the order and spatial relationship they will have in the finished product (ibid., p. 629).

But sometimes we manipulate environmental resources vicariously, as it were, to deal with problems which do not directly present themselves in our immediate environment. This happens whenever we deal with problems about something that is happening in a remote location, or in the future, or perhaps involves the exploration of counterfactual scenarios. For instance, we arrange miniature armies on a map to figure out the best way to attack the enemy. In this case, our more abstract epistemic goal (i.e. launching a successful attack) is no longer directly linked to the immediate spatial aspects of our environment, thus making it a case of *symbolic* off-loading. Another powerful example of symbolic off-loading is our use of Venn diagrams for the purpose of syllogistic reasoning, which depends on a carefully designed mapping between spatial relations among circles and inferential relations among abstract categories (see Chapter 5).

4. The environment is part of our cognitive system. Once it is recognized that environmental resources can play a vital role in assisting and enhancing our cognitive powers, it is tempting to take an extra step to argue that cognition itself is an activity that can literally be spread out across brain, body, and environment (Greeno & Moore, 1993, p. 49; Thelen & Smith, 1994, p. 17; Beer, 1995, pp. 182-3; Wertsch, 1998, p. 518; Hutchins, 1995a; Clark, 1997, pp. 513-6). According to proponents of the extended mind thesis, doing justice to the full interactive complexity of distributed cognition requires that we treat the embo-

8 For a more philosophical discussion of this idea in the context of HEC, see Chapter 2.

died organism *together with* the relevant aspects of its environment as a single cognitive system. Exploring this presumably most radical aspect of situated cognition, which has been dubbed the extended mind thesis, is the principal topic of this book.

5. Cognition is for action. An emphasis on the embodiment of cognition is sometimes construed as the claim that cognition must be primarily understood in terms of its function to control the behavior of an organism. For instance, Clark (1998c, p. 506) suggests that "[b]iological brains are first and foremost the control systems for biological bodies. Biological bodies move and act in rich real-world surroundings." To elevate this claim beyond the rather uncontroversial observation that our minds have evolved for their function to control our bodies, advocates of embodied cognition aim to show that adopting such a perspective yields deeper insights into the structure and current functioning of our cognitive system – insights about our cognitive machinery that might otherwise easily have been overlooked. Wilson (ibid.) cites recent discoveries about the mechanisms of perception (Churchland, Ramachandran, & Sejnowski, 1994; Ballard, 1996; Grafton et al., 1997) and memory (Glenberg, 1997) as examples of such an attempted paradigm shift. For instance, traditional cognitive science has often assumed that the primary purpose of visual perception is to furnish us with a detailed map of the environment that accurately represents objective features of the external world (Clark, 1997). Against this objectivist assumption, enactivists (Regan & Noë, 2001; Noë, 2004; Gibson, 1966) have argued that we primarily perceive objects and situations in terms of their functional relevance to us, in particular in terms of our dispositions to actively engage with the world through motor activities. At the same time, neuroscientists have isolated visual pathways in the brain that are selectively responsible for controlling visually guided action (Milner & Goodale, 1995; Goodale & Milner, 2004).

6. Off-line cognition is body-based. In recent years, evidence has been mounting that cognitive mechanisms and structures that originally evolved for perceptual and motor processing have subsequently been recruited to assist in the performance of more abstract cognitive tasks. The representational function of this cognitive re-deployment is to generate comparatively "cheap" off-line simulations of potentially very costly causal interactions with the external environment (Dennett, 1996; Clark & Grush, 1999; Grush, 2004; see also Wilson, 2002, pp. 632-35). Evidence for the embodiment of *off-line* cognition has been found in the domains of mental imagery, working memory, episodic memory, implicit memory, reasoning and problem-solving, concept formation, as well as our understanding of other minds. For instance, the off-line use of modality-specific sensorimotor simulations in working memory (e.g., "thinking aloud") is a brain-bound variant of cognitive off-loading —i.e., cognitive work is off-loaded onto perceptual and motor control loops which are located within the brain. In cognitive linguistics (Langacker, 1987, 1991; Talmy, 2000), it has been argued that the syntactic encoding of grammatical relations (e.g., predicate-argument structure) is semantically driven by complex "image schemas" which rest on our embodied knowledge of the external world. These perceptually grounded image schemas are organized on the basis of domain-general principles (e.g., figure/ground-distinctions).

The upshot of Wilson's taxonomy is that instead of trying to draw a sharp distinction between embodied and situated cognition, it may be more useful to distinguish between *on-line* aspects and *off-line* aspects of embodied/situated cognition (ibid, p. 635). The on-line aspects of embodiment/situatedness concern the need for our cognitive system to cope in real time with the exigencies of our immediate environment, in a way that is bounded by our limited cognitive resources. The off-line aspect of embodiment/situatedness, on the other hand, refers to the potent opportunistic tendency of the human mind to co-opt ambient resources that are within its immediate reach for cognitive tasks which are *not* directly related to features of one's immediate environment. This tendendy includes the creation and manipulation of external representational vehicles (as in the case of symbolic off-loading), but also the recruitment of previously under utilized internal resources (such as when sensorimotor representations are used to run off-line simulations in the brain). As Wilson (ibid.) aptly remarks, whereas the on-line aspects of situatedness portray the mind as serving the body, it is the mind which appropriates the body and parts of the external environment for its off-line cognitive purposes. It is the latter aspect of embodied/situated cognition to which we shall now turn.

1.2.3 "Distributed" approaches to cognition

Viewed from the "connectionist" take on the biological brain as a giant parallel distributed processor (Rumelhart & McClelland, 1986), one would expect that human beings, by and large, exhibit a mixed cognitive profile which Clark (2001a) once described as being "good at frisbee, bad at logic" (ibid, p. 18). On the one hand, we should be naturally good at what are fundamentally pattern-recognition problems such as sensorimotor coordination or face recognition. On the other hand, we should be naturally bad at tasks that require complex sequential reasoning, logical deductions, and long-term planning. As Clark observes, this profile of the human mind is at once "familiar and alien" to us. The reason it seems alien to us is that we somehow regularly seem to overcome the suggested cognitive limitations when we (e.g.) plan our next summer vacations, submit a term paper, or successfully solve a Sudoku puzzle. The janus-faced profile of the human mind raises the question of *how* exactly human beings manage to transcend the fairly limited innate processing capabilities of their biological brains. Clark has proposed an intriguing answer to this problem:

> "Much of advanced cognition, I believe, depends on the operations of the same basic capacity used for on-line, adaptive response, but tuned and applied to the special domain of external and/or artificial cognitive aids – the domain, as I shall say, of *wideware* or *cognitive technology*" (Clark, 2001a, p. 141).

Clark's proposal highlights the monumental impact which *cognitive scaffolding* has played in the evolution of human cognition (Donald, 1991; Dennett, 1995; Sterelny, 2004a). The notion of cognitive scaffolding, first developed in the writings of Soviet psychologist Lev Vygotsky, forms a center piece of constructivist theories of higher learning (Wood, Bruner, & Ross, 1976; Bruner, 1985; Jonassen, Mayes, & McAleese, 1991). Taken broadly, a cognitive scaffold can be any kind of support structure which enables a student to accomplish cognitive tasks which she couldn't quite accomplish if left to her own devices. More precisely, we can distinguish three kinds of tasks with respect to a learner's cognitive potential at a given time (Figure 1): (i) tasks which the student can perform independently; (ii) tasks which the student cannot perform at all, regardless of any help she may receive; and (iii) in between those two extremes, tasks which can only be performed with some outside help – some *scaffold* – that is provided by a teacher, often in tandem with the right artifact.

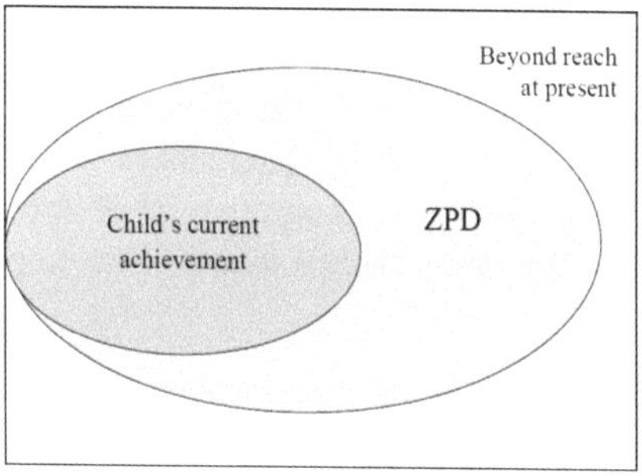

Figure 1. Vygotsky's notion of a "Zone of Proximal Development" (ZPD)

The third category of tasks forms what Vygotsky (1934/1986) called the "Zone of Proximal Development," i.e. the cognitive domain in which a student is ready to grow. True to its original meaning, educational psychologists primarily think of cognitive scaffoldings as temporary aids, which are eventually removed once the student has learned to perform the task at hand independently. But if we carefully inspect our cultural practices of scaffolding, we can see that scaffolds frequently remain in place as relatively permanent augmentations of our cognitive apparatus, because they are practically indispensable for the successful performance of many cognitive tasks that would otherwise baffle our "naked" brains.

Clark's hypothesis about the mechanistic underpinnings of (much of) distinctively human cognition owes much to the idea of "distributed cognition," an alternative framework for cognitive science that was first proposed in the mid-1980s most notably by Donald Norman, Edwin Hutchins, Lucy Suchman, and David Kirsh.[9] It departs from the organism-fixated mainstream paradigms of cognitive-psychology in two crucial respects (Hollan et al., 2000). The first principle, which concerns the boundaries of the relevant unit of cognitive analysis, is that distributed cognitive processes "are delimited by the functional relationships among the elements that participate in it, rather than by the spatial co-location of the elements" (ibid, p. 175). The second principle asserts that the range of mechanisms by which cognitive processes are carried out need not be bound by the skin or the skull of the biological individual. Because of its deep commitment to these principles, the distributed-cognition approach differs greatly from traditional psychology insofar as it extends the realm of what is considered *cognitive* beyond the individual to encompass interactions between people and their tools, as well as interactions between people. From this perspective, human cognition can be distributed

in ***space***, insofar as environmental (natural or technological) structures transform the nature of the cognitive tasks which our biological brains and bodies have to perform;

over ***time***, insofar as the outcomes of earlier stages of cognitive processing transform the task demands during the later stages;

over ***people***, insofar as the division of cognitive labor in a group transforms the nature of the tasks which an individual member has to perform (cf. Hollan et al. 2000, p. 176).

The distributed-cognition approach seeks to integrate methodologies from a variety of disciplines including cognitive science, anthropology, sociology, and human-machine interaction. From cognitive science, it borrows the notion of cognition as computation, understood broadly as the "creation, transformation, and propagation of representational states" (Hutchins, 1995a, p. 50) across different media, whereby it is important that these "media may be inside as well as outside the individuals involved" (ibid., p. 373). But in order to correct for the

9 See especially the papers and books by Norman (1988, 1991, 1993), Zhang & Norman (1994), Hutchins (1995a, 1995b), Suchman (1987), Kirsh & Maglio (1994), and Kirsh (1995, 2006, 2009); for overviews, see Rogers & Ellis (1994), Hollan, Hutchins, & Kirsh (2000), Perry (2003), and Harnad & Dror (2008).

individualist bias which haunts much of traditional cognitive psychology, it also takes on board theories of intersubjective behavior, of spoken, written, and non-verbal communication, social mechanisms of group coordination and the formation of shared intentions and mutual expectations, the division of labor, and organizational learning – all with the intent of incorporating these complementary insights into a general framework for understanding the distributed character of (especially human) cognition. This integrative approach has been successfully used, among other things, to analyze the socially and technologically distributed cognition of navigation crews aboard navy cruisers (Hutchins, 1995a), piloting crews for commercial airliners (Hutchins, 1995b; Hutchins & Klausen, 1996), help desks in call centers (Ackermann & Halverson, 2000), air traffic control centers (Garbis & Waern, 1999), software teams (Flor & Hutchins, 1992), stage acting in the Elizabethan theatre (Tribble, 2005), and NASA's design and execution of the six *Apollo* moon landings (Mindell, 2008).

In an interesting discussion of the role of (what he calls) "epistemic engineering" for human cognition, Kim Sterelny (2004b) has provided his own taxonomy of the myriad ways in which human beings distribute cognition. A brief review of his taxonomy, including a sample of his illuminating examples, will remind us of the central role which scaffolding plays in our everyday cognitive endeavours.

1. We alter our environment to ease memory burdens. Using the external environment to store information is arguably the most basic and fundamental off-loading strategy. First, we often use the environment to cue our memories. For instance, after pouring the last drop of cream into our afternoon coffee, we strategically leave the empty container on the table to remind us of buying new cream the next time we go to the supermarket. When we blaze a new trail through unchartered territory, we make easily-recognizable alterations to the environment so that we (and others) can more easily recall the original path. Second, we create graphical or linguistic records of memorable events in a public medium that we can share with others. Our modern environments brim over with purposefully designed external memory aids that we use to plan our daily activities. Third, before the advent of written language, human beings have long relied on the properties of speech to re-code memorable information in forms that are easier to recall, e.g., in songs, stories, and rhyme. Fourth, establishing a social division of cognitive labor has helped to circumvent individual memory burdens, because individual group members can specialize in different aspects of a collectively shared memory task.

2. We transform difficult perceptual problems into easier ones. When you read these lines, you will notice that I have deliberately highlighted the first and maximally informative sentence of each paragraph, so that it visibly stands out from the rest. Warning signs are made in bright contrasting colors so they capture our attention. Before a carver sculpts a piece of wood with his chisel and hammer, she first marks the anticipated point of contact in order to pinpoint her attention to the working surface. To prepare the assembly a jigsaw puzzle, we first group together all roughly similar-looking pieces, before we attend to more fine-grained similarity classes such as all green pieces with a straight edge (Kirsh, 1995). Quantitative information that is difficult to grasp in verbal form is much easier to process when it is presented spatially in the form of pie-charts, graphs, and maps. Using different words for similar-looking but different things prompts a learner to search for subtle perceptual differences that would allow her to tell one from the other.

3. We transform difficult cognitive problems into easier perceptual problems. A commonly used strategy in scaffolding is to exploit the structure of physical space around us (Kirsh, 1995), because this allows us to re-deploy our perceptual faculties in the service of cognitive tasks. A compelling instance of cognitive off-loading is the much-cited demonstration of "intelligence in the cottage cheese" (De la Rocha, 1985). Preparing lunch, a person who was on a diet had to figure out how much two-thirds of the recommended daily allotment of cottage cheese (= three-quarters of a cup) would amount to. To solve this problem, the person could have converted the relevant instruction into fractions and multiplied 2/3 by ¾. But for the average person, this is often a fairly effortful and potentially error-prone task – even with the help of pen and paper. Instead, the dieter was observed to turn over a one-cup tub of cottage cheese on a plate, criss-cross it with a knife to mark four quarters, remove one quarter, and then eat two of the remaining three quarters – which is of course exactly the suggested one-half.

Another striking example is the transformative role which the process of sketching plays in the creation of abstract art (Van Leeuwen et al., 1999; discussed in Clark, 2001a, pp. 147-50). The reason why artists so heavily rely on their sketchpad is that our "off-line," i.e. imaginary use of visual imagery is constrained in ways in which "on-line" perception is not. Experimental evidence (Chambers & Reisberg, 1985) suggests that while we are quite good at synthetically combining imagined elements into new wholes, our brains are severely limited in their analytic capacity to break down imagined shapes into novel component figures (e.g., as it is required to discover the second interpretation of an ambiguous figure). According to Van Leeuwen et al., the iterated trial-and-error process of sketching allows artists to engage their on-line perceptual capacities to overcome our fixation on established visual patterns in the creation of abstract art.

4. We transform difficult learning problems into easier ones, by systematically changing the information structure of the environment to which the next generation of learners is subsequently exposed. This can be a side-effect of other cognitive activities that are useful in their own right. For instance, the availability of different linguistic labels for similar-looking things serves as an important cue for learners to search for underlying categorical differences. In many cases, we also actively facilitate this process by teaching our offspring how to tell apart an X from a Y. In general, humans invest a lot in making sure that valuable skills and information are reliably passed on from one generation to the next: we deliberately demonstrate skills in a form that makes them easier to imitate (e.g., because they are slowed down, exaggerated, or repeated); we try to teach things in the right sequence so that students can progress from the simple to the more difficult; we provide ample supervision and correct faulty performances; and we use models and artifacts as external teaching aids.

5. We engineer workspaces so that frequent tasks can be completed more rapidly and reliably. Effectively organized workspaces structure our daily routines from relatively mundane activities to the performance of highly specialized but recurring tasks. A guiding principle of intelligent workspace design is to support the distribution of cognitive workload across space and time (Hutchins, 1995a). For instance, cooks often get out and group their ingredients in advance of cooking so they are quickly available when needed, and in a spatial arrangement that matches the anticipated sequence of operations prescribed in the recipe (Kirsh, 1995). Or, in the noisy and hectic environment of a bar, expert bartenders skillfully rely on a physical line-up of distinctively shaped glasses to recall the type of order which they have to prepare, and the sequence in which these orders have been placed (Clark, 2001a, p. 141). As Clark points out, the waiters' reliance on the external scaffold was confirmed by experimental studies in which the shape of the glassware was systematically manipulated (Beach, 1988). In the condition where only identically shaped glasses were allowed, the experts' performance was drastically reduced, whereas the performance of novice bartenders was not significantly affected.

Finally, consider the "three-minute-rule" on which experienced navigators used to rely to simplify the otherwise difficult calculation of distance-rate-time problems (Hutchins, 1995a, pp. 151f). To compute a ship's speed in nautical miles per hour (= 2000 yards), it was a standard navigational practice to take a position fix every three minutes (= 1/20 of an hour), because this means that the distance covered since the previous position fix, measured in hundreds of yards (= 1/20s of a nautical mile), would directly correspond to the speed of the ship in nautical miles per hour. For instance, a ship that has traveled 1500 yards during the last three minutes has a speed of 15 nautical miles per hour. Here, an essential part of the computation consists in the appropriate temporal distribution of cognitive labor.

1.3 Cognitive scaffolding works on multiple scales

The notion of cognitive scaffolding can be generalized, and used to describe a variety of cognition-enhancing agent-environment interactions that occur on different spatial and temporal scales: the phylogenetic evolution of cognition, the ontogenetic development of cognition, and the performance of cognitive tasks here-and-now. Let me briefly say a few words about each type of cognitive scaffolding, and how it relates to the extended mind thesis.

1. Scaffolding cognitive evolution

What makes our minds so markedly different from those of our closest biological kin such as primates and other (now extinct) hominids with similar-sized brains? Which parts of our cognitive apparatus support the so-called "advanced" and seemingly "dis-embodied" modes of problem-solving (e.g., long-term planning, deductive reasoning, and mathematical cognition) that set us apart from our biological ancestors, and that were so highly prized by classical symbol-manipulation approaches in cognitive science? From a bird's-eye perspective, we can distinguish in principle three possible accounts of the evolutionary explosion of intelligence in the human lineage (Clark, 2001a, pp. 140f; 2001b, 2003).

According to the first account, advanced human cognition is the end product of a series of qualitative upgrades that evolution has genetically hardwired into the basic cognitive architecture of our human brains. For instance, evolutionary psychologists tend to view the human mind as a bundle of functionally specialized cognitive modules which are adaptive solutions to specific types of information-processing problems that our ancestors regularly encountered over long stretches of evolutionary time (Barkow, Cosmides, & Tooby, 1992; Pinker, 1997; Buss, 2004). Put in very general terms, the leitmotif behind such a view is to explain the cognitive *discontinuities* between humans and their closest kin directly in terms of underlying biological *discontinuities*.

According to the second account, there is nothing special about the cognitive mechanisms which are responsible for (supposedly) distinctively human modes of thinking. Those mechanisms are either identical with, or at least deeply continuous with the familiar bundle of evolutionarily more basic mechanisms that support embodied/embedded cognition. This kind of "biological cognitive incrementalism" (Clark, 2001a, p. 7) is driven by two related motives. First, it strongly doubts the putative centrality of symbolic, rationative modes of cognition as paradigmatic manifestations of the human mind, and instead places a heavy emphasis on the "on-line" aspects of embodied/embedded cognition (see Chapter 1.2.2). Second, it reflects the empirical success which psychologists have had in applying the theoretical vocabulary, analytic methods, and methodological toolkits borrowed from dynamical systems theory and artificial life to study in much greater detail the embodied aspects of cognition (cf. Clark, 1997 for a review; see also Chapter 3). Taken together, those two motives have nourished the incrementalist belief that a strictly "bottom-up" approach to intelligent behavior will eventually scale-up to human-level rationality by a sequence of relatively minor tweaks of the same old cognitive architecture which powers the embodied/embedded mind. In short, the leitmotif of biological cognitive incrementalism is to portray the deep cognitive *continuity* between human and non-human minds as a result of their fundamental biological *continuity*.

According to a third, "tool-centered" account, the effects of scaffolding have played a most pivotal role for the cognitive explosion in our lineage. It represents an interesting blend of the first two options insofar as it tries to reconcile a deep biological *continuity* with an equally profound cognitive *discontinuity* between humans and their closest biological kin. Unlike any other creature on the evolutionary ladder, we strategically surround ourselves with multiple layers of cognitive scaffoldings that have continually re-shaped and expanded the space of human reason. Thanks to a co-evolutionary spiral of increasingly potent interactions between our brains and our cognitively engineered environments, the cognitive capabilities of enculturated *minds* significantly outstrip the cognitive capabilities of biological *brains*. Different variants of the third account have been defended by Dennett (1995, 1996, 2000), Mithen (1996, 2000), Rowlands (1999), Deacon (1997), Millikan (1993), Donald (1991), Sterelny (2004a), and Clark (2001a) himself.

The scaffolding account immediately raises a "chicken and egg" problem which Clark (2001a, p. 150) called the "paradox of active stupidity": how could the suggested co-evolutionary spiral ever have gotten underway, were it not for the existence of smart brains which are inherently capable of scaffolding their environments, including the creation of smart tools? As we shall discuss in greater detail below (Chapter 4), Clark's solution to this problem lies in a process of iterated bootstrapping. For an explanation of how the process got underway, the scaffolding thesis does not deny that there weren't *any* biologically based differences between the brains (and bodies!) of early humans compared to those of other hominids. Those differences, which at first may have seemed relatively small and insignificant, would still have been enough for human beings to create the first generation of cognitive artifacts. Over evolutionary time, the growing availability and impact of those resources on our cognitive development would slowly lead to an increase in intelligence, which in turn would allow human beings to devise even smarter cognitive tools, and so forth. In short, small, biologically triggered differences would have acted as "seeds" (ibid., p. 151) to initiate a runaway process between humans and their cognitive artifacts.

As a possible short-term illustration of cognitive scaffolding on a cultural time-scale, consider the so-called "Flynn effect." It refers to the widely documented phenomenon that over the past century, the results of intelligence tests have increased at a rate of about 3-5 points per decade (Flynn, 1984). The same basic pattern of increase has been robustly observed, although the exact IQ gains apparently vary with respect to different countries and the type of intelligence test administered (Flynn, 1994). But as a general trend, the generational increases in IQ lie between 5 and 25 points. Notably, however, intelligence has not increased across the board. The gains are most striking for tests which emphasize the ability to recognize abstract, non-verbal patterns, but are far more modest (or even non-existent) for tests that emphasize the kind of "formalized" knowledge (e.g., vocabulary, math skills, factual knowledge) which is taught in school. This would seem to indicate that the Flynn-effect reflects an enhanced aptitude for abstract problem-solving rather than an increase in general intelligence (Dickens & Flynn, 2001).

Dickens and Flynn (2001) have recently proposed a type of "scaffolding" explanation for the Flynn Effect. They suggest that the changing working conditions in industrialized societies bring about a number of environmental changes which would act as potential "seeds" of the Flynn effect. First, since the intellectual demands of an average person's job have steadily increased during the last century, more people seek out higher educational credentials to improve their job opportunities. Second, having more time off work means that people also tend to have more time to devote to cognitively stimulating leisurely activities (such as reading, solving puzzles, gaming), at least if they are encouraged to do so by their social environment. This possible knock-on effect is likely to be amplified by the smaller average size of families, which affords parents more time and better chances to cater to their children's specific intellectual interests. Third, the growing sophistication of cognitive scaffolds (e.g., literacy, numeracy, electronic appliances, digital computers) in which we immerse ourselves daily, and from an early age, might further hone people's test-related problem-solving skills. Even if the direct effects of any one seed would initially not amount to much, their combined effects are reinforced by powerful individual and social multipliers which together might significantly improve the average person's ability to receive higher scores in specific sections of IQ tests.

2. Scaffolding cognitive development

What may have been the single most consequential, biologically triggered change in our brains that jumpstarted the rapid cognitive evolution in our species? For Clark (2001a), the ultimate "seed" is not to be found in the addition of any particular "magic bullet" in our brains. It rather consisted in a general design innovation towards a hitherto unparalleled level of neural plasticity, together with a concomitant extension of our critical learning period ("childhood"). This biological change opened an extended window of opportunity for our scaffoldings to make a deep and lasting impact on the ontogenic development of our brains, a thesis sometimes referred to as "neural constructivism" (Quartz & Sejnowksi, 1997; Quartz, 1999; Baltes, Reuter-Lorenz, & Rösler, 2006; Mareschal et al., 2007a, 2007b). Neural constructivism is a natural ally for anyone who – like Clark – wants to tell a tool-centered, co-evolutionary story about the evolution of human intelligence, because it offers the promise of a neural mechanism by which artifact-rich environments reliably sculpt, enhance, and transform our cognitive regimes.

The most recent evolutionary additions to the human brain – the neocortex, especially the prefrontal lobes – exhibit a profound developmental plasticity which is particularly well-adapted to extract statistical regularities from the world that surrounds us. There is considerable evidence that cerebral growth in the human cortex is surprisingly experience-dependent, down to the stimulus-driven construction of dedicated neural circuitry (Baltes et al., 2006; Mareschal et al., 2007a; 2007b). This suggests that culturally enhanced learning does not just lead to an accumulation of more knowledge, but can significantly alter the internal architecture of the learning device itself. In support of neural constructivism, Clark (2001a) cites a variety of studies: first, cortical transplantation experiments in which neural tissue from the visual cortex was injected into other locations (e.g., somatosensory, auditory cortex), and was plastic enough to acquire the functional response characteristics of its target location (Schlagger & O'Leary, 1991; Roe et al., 1990); second, evidence of how the acquisition of specific cortical response patterns depends on prior developmental interactions between given parts of the cortex and specific input signals (Chenn, 1997); third, computational models that show how artificial neural networks can alter their architecture (e.g., number of layers, unit) over time in response to learning specific input patterns (Quartz & Sejnowski, 1997); fourth, we can add developmental systems theory as an account of how our prolonged period of childhood provides an ecologically extended mechanism by which artifactual, linguistic, and socio-cultural resources scaffold human developmental processes (Griffiths & Stotz, 2000; Jablonka & Lamb, 2005).

3. Scaffolding cognitive processing

The third aspect of scaffolding concerns the active-causal role of environmental structures in assisting and structuring our concrete performance of cognitive tasks. If we look at the recent literature, there are two subtly different ways of articulating the cognitive benefits accrued from scaffolding (for more on this, see Chapter 4). One line of thought which is most strongly associated with the work of Daniel Dennett (e.g., 1995, 1996, 2000) emphasizes the developmental effects of scaffolding on the neural circuitry of our brains. At the center of this perspective, which has distinctively Vygotskian roots, is the portrayal of cognitive development as the internalization of socio-cultural practices which are at first explored in a public domain (e.g., language, writing systems). Andy Clark's (2003, 2007, 2008) emphasis, on the other hand, has been on our continual reliance on bio-external structures for the success of distributed problem-solving and information-processing ensembles, understood as essentially hybrid, but tightly integrated cognitive systems that span across neural, bodily, and environmental resources. The functional integration of biological and technological resources, of nature and nurture, is adroitly described by Clark's (2001a, p. 151) metaphor of "cognitive dovetailing." But how exactly should we understand this enticing idea?

If we skim through the more philosophical treatments of the DEEDS-literature, the language used to characterize cognition as a "distributed" or "extended" phenomenon often takes a notably flamboyant turn. For instance, commenting on the adaptive value of cognitive scaffolding, Dennett emphasizes that a primary source of our greater intelligence is "our habit of off-loading as much as possible of our cognitive tasks into the environment itself – *extruding our minds* (that is, our mental projects and activities) into the surrounding world" (Dennett, 1995, p. 134; my emphasis). With similar pizzaz, John Haugeland (1998) claims – after echoing Dreyfus' reminder that "intelligence itself abides 'out' in the world, not just 'inside'" (ibid., p. 232), that "much as an internal map or program, learned or stored in memory, would […] have to be deemed *part of* an intelligent system that used to get it to San Jose, so I suggest the *road* should be considered *integral* to my ability" (ibid., p. 234). In their seminal paper on the "extended mind," Clark and Chalmers (1998) maintain that "one can explain my choice of words in Scrabble, for example, as the outcome of an extended cognitive process involving the re-arrangement of tiles on my tray. […] In a very real sense, the re-arrangement of tiles on the tray is not part of action; it is part of *thought*" (Clark & Chalmers, 1998, p. 10). And after illustrating how much our brain relies on external resources to write an academic research paper, Clark concludes that "it is this larger system which solves the problem. […] The intelligent process just *is* the spatially and temporally extended one which zig-zags between brain, body, and the world" (Clark, 2001b, p. 132).

What claim do these authors wish to make with their often metaphorical language? What are the "ontological commitments" of the extended mind thesis? Those are the kinds of questions which take us right into the heart of the debate which is the central topic of this book.

1.4 Is cognition extended or "merely" embedded?

The most literal – and thus most radical – way of understanding the above formulations is the extended mind thesis (Chapter 1.1; cf. Clark & Chalmers, 1998; Clark, 2003, 2008). It is thus time to cast the thesis into a canonical form that we shall call – following Rupert (2004, p. 389) – the "hypothesis of extended cognition":

HYPOTHESIS OF EXTENDED COGNITION (HEC): A significant portion of human cognition is not confined to the boundaries of our brains, but literally includes bodily and environmental resources among its (mereological) parts.

Put succinctly, one could simply say – echoing the words of Hilary Putnam (1975b) – that cognition ain't (just) in the head. Still, it is an undeniable conceptual virtue of the somewhat stilted formulation of HEC that it nicely accentuates a number of features which demand our philosophical attention. Let me briefly mention a few.

First, recall our earlier distinction between two varieties of psychological externalism, each of which contradicts a different ingredient of Cartesian internalism – either the possession-claim or the location-claim (Chapter 1.1). Framed as a thesis about the *how* and *where* of cognition – i.e., the mechanistic nuts and bolts of cognitive processing – the HEC is in direct conflict with the location-claim of Cartesian internalism. Its truth does not logically depend on, nor does it imply the endorsement of content-externalism (cf. Chapter 2.10).

Second, HEC is not merely an assertion of the metaphysical (or indeed nomological) possibility of extended cognition. It is presented by its main proponents and critics alike as a theoretical hypothesis in cognitive science that purports to capture a central and distinctive trait of human cognition. Even if the careful analysis of a single thought experiment may in principle suffice to show the metaphysical possibility of isolated episodes of extended cognition, we agree with the critics that "HEC cannot be vindicated by oddball cases, science-fiction thought experiments, or the development of assistive technologies" (Rupert, 2010, p. 343). To make a substantial difference to the causal-explanatory practices of cognitive science, proponents of the HEC ought to produce evidence for the theoretical fecundity of their framework. This can be achieved in a variety of ways, e.g., by inspiring the development of new research methodologies, by suggesting innovative experimental paradigms, but also by providing a more unified and attractive theoretical gloss on existing data (cf. Rupert 2009, p. 18; Clark 2008, pp. 218f).

Third, it is worth pointing out that the HEC is couched in terms of the more technical notion of *cognition*, rather then the more vernacular, but philosophically "loaded" notion of the *mental*. But what is meant by cognition? Consider, first, the sheer heterogeneity of psychological predicates that one may want to include in the overarching category of mental or cognitive properties. From a folk-psychological perspective, we attribute paradigmatic mental states like beliefs, desires, intentions, and consciousness to cognitive systems. Going along with these are capacities more typically studied in psychology labs such as perception, attention, motor control, memory, learning, categorization, inference, planning, or creativity. From a more abstract perspective, cognitive science has tried to characterize cognition in terms of computation, information processing, the vocabulary of dynamical systems theory, the generation and use of representations, and problem-solving flexibility. It makes good sense to call all these properties cognitive, based on a firm but defeasible pre-theoretical understanding of the distinctive roles which they play in the production of intelligent behavior. Such an understanding may be presumed innocent until proven guilty. However, even though there are certainly important family resemblances among those properties which allow us to group them together under a single umbrella, we do not have to assume that there must be a universally shared, theoretically deep feature – a "hidden essence" – of the cognitive for such a categorization to be useful. This caveat should curb our enthusiasm for the critics' project to defeat the HEC by finding the putative "mark of the cognitive" (Adams & Aizawa, 2001, 2008), or by providing a general "demarcation criterion" (Rupert, 2009) on the basis of which one could exhaustively divide the world into cognitive versus non-cognitive resources (cf. Wilson, 2010c). A more realistic, and ultimately more fruitful approach to the HEC, as we shall see, is to break down the generic notion of cognition into a number of more specific traits, all of which have been proposed as hallmark features of cognition, and then to defend (or reject) the plausibility of the HEC for each specific construal of cognition under consideration.

We should point out, though, that there is a potential lacuna between the more liberal notion of a *cognitive* state (or process) and philosophically more "loaded" construals of a *mental* state. Suppose we require – as many philosophers do – that full-fledged mental states (such as beliefs or desires) must have a special kind of representational content that is roughly propositional in nature, unlike "sub-doxastic" (Stich, 1978) cognitive states that are causally implicated in cognitive processes (e.g., visual recognition, spatial cognition) without having propositional content. This leaves open the possibility that cognitive processes are extended, whereas intentional mental states are not. Clark and Chalmers (1998) anticipate this potential limitation of their thesis, and thus "[...] propose to take things a step further. [...] In particular, we will argue that *beliefs* can be constituted partly by features of the environment when those features play the right sort of role in driving cognitive processes. If so, the mind extends into the world" (ibid, p. 12). My point here is not (yet) to defend their conclusion, but merely to draw attention to the fact that HEC *may* turn out be true with respect to some construals of cognition, but false for others. In particular, I shall officially remain agnostic on the question of whether *conscious* mental states extend into the environment – a claim over which both Clark (2009) and Chalmers (2008, pp. x-xi) have recently expressed strong reservations. As I hope to show, there is enough room to defend a robust version of HEC while retaining the idea that phenomenally conscious states of our minds supervene on our organism-bound neural machinery (see also Chapter 1.5).

Fourth, I prefer the canonical formulation of HEC over Clark and Chalmers' grand claim that "the mind extends into the world" (1998, p. 12). In order to demonstrate that parts of the mind literally extend into the world, one ought to have a firm theoretical grip on what it means to be a part of someone's mind. In order to avoid this tricky subject, I suggest that we translate the locution of an "extended mind" into the HEC along the following lines. (1) A mind is an entity that can have mental states, capacities, and engage in mental processes (which can be understood as causally ordered transitions between mental states). (2) A mind is a complex physical entity that is entirely composed of physical parts. The physical realization of mental states (at a given time) is constituted by the appropriate organization of those parts. (3) According to the HEC, some mental states or processes are physically realized partly in virtue of extra-neural resources that are part of an organism's body or external environment. (4) A mind is located wherever its mental states are physically realized.[10] This is certainly only an approximate translation procedure, but I won't try to tighten it up until later when we'll discuss some further metaphysical complications which arise from the notion of physical realization (Chapter 6). For now, suffice it to say

10 As I will argue later (Chapter 6), claim (4) is not strictly speaking required for a defense of HEC (Chapter 6). As Rob Wilson (2001, 2004) has argued, mental or processes can be extended because they have wide realizations whose extension goes beyond the organismic boundaries of their proprietary bearers ("subjects"). If we identify minds with the bearers of mental properties, this means that there can be extended mental states without extended minds.

that establishing the HEC is enough to defeat the location-claim of Cartesian internalism.

Now, provided that we grant all these concessions and distinctions, where's the (philosophical) beef? What is it about the HEC which has struck some commentators as "deeply problematic" (M. Wilson, 2002, p. 625)? There is one big issue which has garnered most of the attention in the current philosophical debate over HEC. Suppose it is true, critics have argued, that a considerable amount of our cognitive practices is surprisingly dependent on epistemically potent forms of bodily actions, and the canny exploitation of environmental props and scaffolding. But why should we infer from this that cognition itself is literally extended? In other words, why should we think that the suggested dependence of the brain on the body and its environment is *constitutive* of cognition, rather than reflect a merely *causal* or *instrumental* dependence? Why isn't it enough to say – what would still seem to be a significant empirical discovery – that our organism-bound cognitive apparatus makes efficient *use* of external, but non-cognitive tools and aids? Why consider the environment as a literal *part* of cognition?

Margaret Wilson (2002, p. 630) has used an interesting analogy from the history of physics to cast doubt on the plausibility of the HEC. She argues that scientists before 1900 already knew a fair amount about the reaction of hydrogen when it interacted with other chemicals. However, it was not until the discovery of the atomic structure of matter that scientists were able to give an explanation of the mechanisms underlying these interactions, and thus to predict how hydrogen would behave in novel situations that had never before been observed. Hence while it is certainly true that the aggregate behavior of hydrogen is always a joint effect of the nature of hydrogen plus its surrounding elements and forces, "explanatory satisfaction came from understanding the workings of the narrowly defined system that is the hydrogen atom" (ibid.).

In similar vein, Fred Adams and Kenneth Aizawa (2010) have argued that the "coupling-constitution fallacy" is "the most common mistake extended mind theorists make" (p. 68) when they tacitly move "from the observation that a process X is in some way causally connected (coupled) to a cognitive process Y to the conclusion that X is part of the cognitive process Y" (ibid., 2009, p. 81; see also Adams & Aizawa, 2008, Chapters 6-7; Adams & Aizawa, 2001). As a non-cognitive example where said fallacy should be apparent, they point out that the expansion and contraction of the bimetallic strip in a thermostat is closely coupled to the ambient temperature of a room, yet this gives us no reason to consider the expanding and contracting strip as extending into the room (Adams & Aizawa, 2009, loc. cit.). A similar criticism is made by Robert Rupert (2004, pp. 395-6), who argues that even if one grants that we cannot fully comprehend many episodes of human cognition unless we consider the wider context in which they occur, it does not follow that the embedding context is therefore part of cognition. For instance, to understand the historical dimension of Nazi Germany's invasion of Poland, one has to know a great deal about the economic conditions in Germany during the 1920s, but this does not make the latter (mereological) parts of the invasion.[11]

In light of those and many other critical considerations, Rupert (2004, 2005, 2006, 2009, 2010a, 2010b) has proposed a competitor perspective to the HEC which he calls the

HYPOTHESIS OF EMBEDDED COGNITION (HEMC): Our cognitive processes occur entirely within the organism, but sometimes depend very heavily, in hitherto unexpected ways, on organismically external props and devices (2004, p. 393).

Rupert's strategy has been to pit the hypothesis of embedded cognition (HEMC) against the hypothesis of extended cognition (HEC), and challenge the HEC on the grounds that the HEMC yields an empirically equivalent, but theoretically more attractive gloss on whatever insights the DEEDS-literature may provide. Adopting the HEC, says Rupert, comes at the expense of abandoning our successful organism-bound methodological practice in psychology, and upsets a variety of independently plausible assumptions about the mind. In order to offset these severe costs, the HEC would have to provide genuine novel insights into the nature of cognition, which however is not the case. In fact, for every HEC-type explanation, we can always provide an empirically equivalent, but theoretically more conservative analysis in terms of the HEMC. Consequently, we should prefer the HEMC over the HEC on broad grounds of ontological parsimony and methodological conservatism. Thus no radical overhaul of our traditional ontology of the mental is necessary to accommodate the effects of cognitive scaffolding and the use of external props and tools to aid our brain-bound cognitive systems.

Rupert (2004, p. 393, fn. 9) instructively compares HEMC-style explanations with a type of cognitive-scientific explanation that Clark (1997, pp. 105ff) dubbed "catch-and-toss"-explanations. Intuitively speaking, the underlying idea here is that the mind/brain "catches" perceptual inputs from the world, and "tosses" back actions as outputs from the mind; but does so in a more interactive manner (e.g., by actively performing bodily movements that are poised to gener-

11 In Chapters 3-4, I shall question the cogency of these supposed counterexamples to the HEC, and argue that the HEC is not guilty of committing any of the suggested fallacies. For now, I simply rehearse these objections as a backdrop for our ensuing discussion.

ate an enhanced range of sensory input) than it is usually recognized by orthodox psychological internalists. As examples of HEMC-style explanations, Rupert (2004, pp. 393ff) refers to the work of Brooks (1997; see also 1999), O'Regan (1992), Churchland et al. (1994), and McClamrock (1995).

Now consider the alternative between HEC and HEMC. Should we treat the two as a theoretical dichotomy, or rather as different poles along a continuum? How do we, or perhaps how should we demarcate the boundaries of cognitive systems, and/or cognitive processes? Many people have strong intuitions that some collections of things (e.g., solar systems, immune systems, economic systems) form a single, unified *system*, while other collections are merely an *aggregate* of haphazardly juxtaposed elements. But on what criteria are those intuitions based, and can we use the same criteria to delineate the boundaries of the mind? On a first and very rough pass, we could say that we individuate *systems* in terms of (i) their *organization*, which can be defined in terms of spatial, temporal, causal, and functional relations between things that are said to be part of the system; (ii) the global *function* which we ascribe to the system as a whole; (iii) the requirement that the properties of things which form elements of a system are somehow *causally affected by their participation* in the system.

For instance, if we look at our bodies, many of our corporeally bound organ systems uncontroversially satisfy these requirements. For instance, physiologists consider the heart and blood vessels as parts of the human cardiovascular system, a system whose boundaries are defined by its global function to distribute blood through the body. Each part of the system makes some distinctive contribution to the function that is performed by the system as a whole: the role of the heart is to pump blood through the body, the arteries' role is to take blood from the heart and supply it to various tissues via the capillaries, and the role of veins is to collect deoxygenated blood from the tissue and return it to the heart (and ultimately to the lung, where the blood is oxygenated). On this count, lymph nodes and lymph vessels do not form parts of the cardiovascular system, because they play no direct role for the contribution of blood. Their function is to distribute lymph, not blood. On the other hand, if we adopt a wider definition of distribution networks in the body, both the cardiovascular and the lymphatic system can be seen as parts of the more encompassing circulatory system. As this simple example shows, what constitutes the boundaries of a system critically depends on how we choose to *individuate* the system (especially its function) as a whole with respect to its function, which in turn depends to some extent on our explanatory interests.

However, it is important not to overstate the suggested context-sensitivity of individuating biological and other kinds of systems. In biology, there are numerous fairly uncontroversial examples of trans-corporeal organ systems, i.e.

biological systems whose functional organization criss-crosses the boundary between organism and environment (see also Chapter 2.9). In species which employ coprophagia, like rabbits and guinea pigs, the digestive system is not bound by the skin of the organism, because the ruminant parts of their digestive systems alone would not be capable of decomposing enough of their ingested plant matter to extract the nutritional value which is necessary for survival. As the ingested food makes its first pass through their intestines, it is infiltrated with bacteria whose digestive enzymes help to break down hitherto undigested materials, before it is excreted in the form of soft caecal pellets. In order to give this partially digested food another pass through their guts so it can eventually be absorbed into the body, coprophageous animals then re-consume these caecal pellets after excretion.[12] Similarly, because spiders can only eat their food in liquid form, some spiders inject saliva with digestive enzymes directly into the body of their prey to liquify its inner organs and tissues, so they can literally suck it up later. As part of this "extra-oral" method of digestion, the prey serves as a kind of external "stomach" for the spider where the extraction of nutrients occurs. In these biological examples, the digestive system is literally extended: it "loops out" into the environment to perform its biological function.

So if we accept that there are extended mechanisms of digestion, do we have similar reasons to believe that cognitive systems might be extended at least in some species?[13] And most importantly, are we human beings one of those species? In this book, I shall argue that we are. Contrary to what Descartes believed, the advanced human cognizer is essentially an extended cognizer.

1.5 Three ground rules for our debate

To manage our discussion over the bounds of cognition, it is essential that we first establish some rules of engagement. In what follows, I propose to adhere to three maxims of philosophical methodology which I call the principles of FAIRNESS, CANDOR, and EXTENDED FAIRNESS. As it is customary, let me review the content of our ground rules so that the real debate may commence. My formulation of these rules will be brief and somewhat fuzzy, but it is better to enforce fuzzy ground rules than not to enforce any ground rules at all.

First, we need a principle regarding the status of our debate within the larger edifice of philosophical inquiry. Philosophical problems have a quaint tendency to get out of hand rather quickly. For instance, we said that in order to make

---

12 See the Wikipedia entry: http://en.wikipedia.org/wiki/Coprophagia.

13 Interestingly enough, Adams and Aizawa (2001, p. 46f), who are two of the staunchest critics of the HEC, readily acknowledge that digestion is a transcorporeal process in coprophagous animals.

sense of the notion of an extended mind, we have to presuppose that minds are composite entities that are located wherever its physical parts are located. But not everybody agrees that minds in this sense could truly exist. In a famous paper titled "I do not exist," Peter Unger (1979) uses the classical paradox of the heap to argue for mereological nihilism, i.e. the radical philosophical thesis that objects with proper parts do not exist. Hence the composite physical entity that one might think of as the referent of the name "Peter Unger" does not exist. If Unger is right, doesn't this rule out the existence of extended minds right away? Perhaps, but deriving the falsity of HEC from the truth of mereological nihilism would amount to a "win by default" – a bogus win that was earned only because one's opponent failed to show up. I take it that this would be an unsatisfactory outcome. Here as everywhere else, picking the right contrast class matters. We wouldn't be shocked to learn that psychological externalism is false because mereological nihilism is true. We'd rather want to know whether psychological externalism is true because psychological internalism is false (or vice versa). To restrict the scope of our discussion (but without begging the question one way or the other), we need a rule about what we can safely consider as "default assumptions" in the context of our debate. Such default assumptions can't be challenged by internalists and externalists alike. Surely the existence of composite material objects ought to be one of those assumptions in the present context. I shall thus ban from our debate anybody attempting to deliver knock-out punches of the Unger-type by adopting the maxim of FAIRNESS:

RULE #1: FAIRNESS. Don't burden the mind with (almost) everything else's problems.

The rule of FAIRNESS has been advocated in the philosophy of mind by Georges Rey (1997), who provides it with the following gloss: "Always ask whether a problem is *peculiar* to the mind, or whether the issue could equally well be raised in other less problematic areas. If it can be, settle it for those areas first, and then assess the philosophy of mind" (ibid, p. 27). FAIRNESS might be construed as a rain check to remain neutral with respect to metaphysical questions of global significance. This is certainly not what I have in mind, since I think that the aim of ontological neutrality is not only illusionary but also seriously misguided. Any claim about the nature of mind and cognition engenders substantive ontological commitments that we ought to flesh out, and "naturalized" metaphysics (in the sense in which it is understood here, see below) is very much in the business of doing that. My reason for adopting the rule of FAIRNESS is, on the other hand, to discourage philosophers from defending idiosyncratic theses about the mind on the basis of supposedly intuitive ontological assumptions which turn out to be implausible once we try to generalize them to other, non-mental aspects of the world.[14]

Second, since this book is about the existence of extended minds, we need a principle by which we adjudicate questions of *existence*. For a "naturalized" philosopher like the author of this book, science is a very reliable (although not necessarily the only) place to look in order to determine what there is. This underlying motive of philosophical naturalism is often associated with Quine's project to make philosophical inquiry continuous with the scientific enterprise, in that it views philosophy in a way that is neither logically prior nor epistemologically privileged over science. It also arises from a healthy dose of respect for our scientific methodologies, based on their impressive track record to ask— and, more importantly, to answer— penetrating questions about the nature of things. As a way of settling questions about existence, Quine (1953) suggested that we ought to be "ontologically committed" to all and only those entities over which we quantify in a logically regimented form of our best scientific theories. Sharing Quine's respect for scientific inquiry, but remaining agnostic on the question of whether the success story of science automatically rules out other ways of determining what exists, I propose as my second rule the principle of CANDOR:

RULE #2: CANDOR. We ought to be ontologically committed to all entities which are indispensable to our best scientific theories.

Put in slogan form, CANDOR implies that the mind is extended just in case our best sciences of the mind turn out to be extended. It is no easy task to say what makes the reference to a given entity "indispensable" for a scientific theory. On a first pass, we can say that an entity is indispensable if its omission from a given scientific theory makes the resulting theory less *attractive*. I won't try to spell out in full generality what that entails, since the attractiveness of a theory depends crucially on how we decide to weigh different criteria such as its descriptive adequacy, instrumental value, explanatory power, degree of unification, simplicity, conservatism, etc. This implies, on the one hand, that questions of existence are inextricably bound up with the pragmatics of scientific explanation. But on the other hand, since scientists are not necessarily in a privileged position to read off the ontological commitments of their theories, it also secures an important job for us naturalized philosophers to help ascertain the overall attractiveness of a scientific theory vis-à-vis its competitors.

Finally, a third rule is needed to protect the HEC from criticisms based on problems that even our best materialist theories of brain-bound cognition are

---

14 As an example, see our discussion of wide realizations in Chapter 6.

currently unable to solve. Since our understanding of how cognitive activities can arise from a medium-sized blob of brain tissue is still fairly limited, we should expect that there are unanswered questions aplenty. But surely it would be unfair to criticize the HEC for deficits and shortcomings which the HEC shares not only with orthodox brain-bound theories of cognition, but also with its main competitor, the HEMC. Thus I further propose the rule of EXTENDED FAIRNESS:

RULE #3: EXTENDED FAIRNESS. Don't burden the theory of the extended mind with problems that equally affect theories of the non-extended mind.

Any attempted refutation of the HEC which generalizes to the HEMC, or to any other orthodox internalist theory of cognition, violates EXTENDED FAIRNESS. In other words, it is incumbent on the internalist critic of the HEC to show that a materialist, externalist theory of the mind is problematic for reasons that do not affect materialist theories of the mind which are predicated on Cartesian internalism (or HEMC). A brief note about the philosophical debate over functionalism will serve to illustrate the proper use of EXTENDED FAIRNESS.

In the philosophy of mind, functionalism is the claim that "what makes something a mental state of a particular type does not depend on its internal constitution, but rather on the way it functions, or the role it plays, in the system of which it is a part" (Levin, 2010). An important consequence of functionalism – commonly referred to as the so-called *multiple realizability* of mental states – is that any two systems with the same kind of functional organization can in principle instantiate the same types of mental states. Commenting on the substrate-neutrality of mental states in their recent defense of the HEC, Rob Wilson and Andy Clark (2009, p. 69) remind the internalist that the "tenet of computational promiscuity" has long been a fundamental assumption of orthodox cognitive science – i.e., the tenet that "pretty much anything, including any kind of processing or encoding, if it is properly located in some ongoing web of computational activity, can contribute to the nature and unfolding of the cognitively relevant computations, and hence emerge as a proper part of the material substrate of some target cognitive process or activity."

In philosophical circles, the implications of multiple realizability have sometimes been used to chastise functionalism for its unabashed "liberalism," based on arguments that allegedly show that water buckets (Block, 1978), rocks (Putnam, 1988), or walls (Searle, 1992) could equally have mental states. A related critique of functionalism is that it cannot capture the purely qualitative aspects ("qualia") of our conscious experience. This is because experiential states such as perceptions, introspections, emotions, and bodily sensations appear to have an essential perspectivalness of "what's it like" to have them (Nagel, 1974), and this aspect has proved to be recalcitrant to a purely functional characterization (Levine, 1983; Chalmers, 1996). Now suppose a critic attempts the following refutation of the HEC:

1. HEC is logically dependent on, or at least most plausibly understood in terms of some version of (extended) functionalism.
2. Functionalism is a fundamentally flawed theory of mental states, because it cannot in principle account for feature X which is an essential property of some mental states (e.g., phenomenal consciousness).

Therefore, the HEC is false.

I shall rule out arguments of this sort from our debate, because they blatantly violate the principle of EXTENDED FAIRNESS. The very fact that the qualitative nature of mental states is considered to pose an equal challenge for internalist and extended versions of functionalism shows that the problem cannot be specific to the HEC (rather than functionalism *simpliciter*). This is especially relevant for the problem of consciousness, because we have already decided to remain agnostic on the question of whether conscious mental states partly extend into the environment. Since our version of the HEC does not imply extended consciousness, we can safely remain agnostic on the bigger question of whether phenomenally conscious mental states are in principle amenable to a functionalist analysis.

2. A primer on the principle of parity

2.1. Introducing parity

In Chapter 1, we saw how profoundly human cognition relies on our habits of skillful bodily action, and the manipulation of environmental props and tools. This led us to raise the question of whether the fact that human cognition is scaffolded supports the stronger claim that it is literally extended (Chapter 1.5). As the critics have argued, the diligent use of environmental resources – while certainly instrumental for the successful performance of many cognitive activities – does not imply that those resources are constitutive parts of our cognitive apparatus. It takes an extra step to justify the suggested inference from cognition as an embodied and embedded phenomenon to the hypothesis of extended cognition (HEC). In their seminal paper, Clark and Chalmers (1998, p. 8) tried to bridge this inferential gap with a principle that has since become known as the *Parity Principle* (PP). In its original formulation, PP states that:

> "If, as we confront some task, a part of the world functions as a process which, *were it done in the head*, we would have no hesitation of recognizing as part of the cognitive process, then that part of the world *is* (so we claim) part of the cognitive process."

According to Clark & Chalmers' paraphrase (ibid.), the message of PP is that a causal distribution of cognitively relevant labor demands a proportionate spread of cognitive credit. Viewed in this sense, PP is most plausibly understood from a broadly speaking *functionalist* theory of mental states (Putnam 1975a, Fodor 1974, Block, 1996). If mental states are individuated on the basis of their functional role in the production of intelligent behavior, there is no principled metaphysical reason why extra-neural states should not be able to earn their keep as mental states – provided, of course, that they play an appropriate functional role. On a first pass, we can say that PP simply asks that we ought not to discriminate against the cognitive status of bodily or environmental states purely because they are located outside the brain. As Clark (2005a, p. 2) put it, PP seeks to erode an unreflective "bio-chauvinistic prejudice" that unjustifiably favors location over functionality when it comes to demarcating the bounds of cognition.

Clark & Chalmers illustrate their argument from PP with two examples that have become part of the stock-in-trade of the debate over the extended mind. Their first example, in which PP is applied to the domain of cognitive processes, involves a thought experiment that draws on the idea of *epistemic action* as introduced in a classic study by Kirsh and Maglio (1994). To demonstrate how closely cognition is linked to action, Kirsh and Maglio investigated human problem-solving in the context of the video-game Tetris. To succeed in Tetris, a player has to determine under time pressure whether two-dimensional blocks of various shapes, displayed in falling motion on a computer screen, fit into slots in an emerging wall at the bottom of the screen. Traditional information-processing approaches to problem-solving predict that rational players would carry out actions that bring them closer to their desired goal state, which is to maximize the surface area of the emerging wall. However, Kirsh and Maglio found that expert players did not conform to this expectation. They often manipulated the falling tiles in ways that would seem to be disadvantageous and irrational from a traditional perspective. For instance, experts would often rotate blocks on the screen just to see whether they fit into one of the slots. If it doesn't fit, performing this type of exploratory action obviously does not bring the player any closer to her goal. Interestingly, the same type of seemingly "erratic" behavior was observed much less frequently in novice players. Faced with the problem of deciding where to put a falling tile, novice players usually did not perform any overt actions at all, but attended to mental images of the falling blocks which they rotated in their heads. Prima facie, this would be evidence against the assumption that expert players of Tetris are more rational problem-solvers than novices.

In order to explain their observations, Kirsh and Maglio introduced a new category into their information-processing analysis. They argued that in addition to regular physical actions, people also perform *epistemic actions*, i.e., actions they perform not with the primary goal of bringing about any physical changes in the environment, but with the aim of changing their own informational state as problem-solvers. This means that in order to predict a rational agent's behavior, we always have to consider a joint state space that includes physical as well as epistemic goals. What the Tetris experts were so prolific in doing was maximizing the effects of epistemic agency. This is because as a method of solving the spatial reasoning problems encountered in Tetris, rotating tiles directly on the screen turns out to be a faster and less error-prone strategy than rotating mental images of tiles in one's head. From their fresh theoretical perspective, they were able to show that expert players were indeed the more rational problem-solvers.

Kirsh and Maglio offer several, non-identical glosses of the cognitive benefits that people can accrue from epistemic actions – "actions that use the world to improve cognition," "actions [that] are used to change the world in order to simplify the problem-solving task," and "actions performed to uncover information that is hidden or hard to compute mentally" (ibid., p. 2). Despite subtle differences in these formulations, the upshot of their study is that many difficult perceptual and cognitive problems are often solved more quickly, easily, and reliably by performing epistemic actions in the world than by performing the corresponding operations in one's head.

In their argument for the HEC, Clark and Chalmers play off Kirsh and Maglio's characterization of epistemic actions by comparing three ways of Tetris-cognition. Player 1 constructs visual images of the falling blocks in her head, and mentally rotates each image to decide whether it would fit into the socket. Player 2 rotates the falling blocks directly on the screen by pressing a button on the video game console. Player 3 is an interesting blend of the first two. Player 3 can decide whether to use good, old-fashioned slow mental rotation, or to activate a neural implant that allows her to fast-rotate the internally generated mental images with the same speed as the on-screen rotations performed by Player 2, but without having to press any buttons. We can further imagine that the neural implant implements the same algorithm for rotation that is used by the computer, is also initiated by neural activity in the motor cortex, and produces transformed visual images of blocks. Clark and Chalmers argue by comparison that the operations performed by Players 2 and 3 are functionally equivalent to the operations performed by Player 1. Consequently, PP implies that they ought to be regarded as cognitive processes of the same type (i.e., instances of *mental rotation*). Their functional equivalence holds true regardless of whether the process of rotation is controlled manually by pressing a button (Player 2) or remote-controlled by a neural implant (Player 3), and also whether it is performed on representations of objects inside the head (Player 1) or directly on objects in the world (Player 2).

Clark and Chalmers' second thought experiment – the tale of Otto – is constructed to extend the HEC to paradigmatic types of intentional mental states, in particular to beliefs. Otto is an imaginary patient who suffers from a mild form of Alzheimer's disease. To compensate for his impaired biological long-term memory, Otto always carries around in his pocket a notebook where he writes down all valuable pieces of information. When Otto discovers that there is a new exhibition hosted at the Museum of Modern Art (MoMA) in New York, he automatically looks up the information that MoMA is on 53rd Street in his notebook, and walks off into the corresponding direction. Now compare Otto's behavior with that of his good friend Inga, whose biological memory is completely intact. Inga similarly desires to go to MoMA and also believes that it is located on 53rd Street. With respect to their actions of walking to the museum in order to see the exhibition, the content of Otto's notebook is functionally equivalent to the content of Inga's biological memory. For instance, we can suppose that the representations in Otto's notebook causally interact with his other beliefs and desires in much the same way as Inga's memories, and that Otto continuously revises his notebook in accordance with new information. Leaning on PP, Clark and Chalmers argue that the entry in Otto's notebook should count as an extended belief, having the same type of intentional content as Inga's brain-bound belief that MoMA is on 53rd Street.

The cogency of Clark & Chalmers' argument depends crucially on the plausibility of PP, whose exact interpretation has turned out to be a matter of considerable controversy (cf. Rupert, 2004, 2009; Clark, 2005a, 2008; Menary, 2006, 2007; Adams & Aizawa, 2001; 2008, Chap. 8; Wilson and Clark, 2009; Sprevak, 2009; Sutton, 2010; Wheeler, 2010a). My goal in this chapter is to summarize and clarify this debate. To that end, I will elaborate on the inferential role of PP, canvass the most commonly raised criticisms of PP, and evaluate the responses which have been offered on its behalf.

2.2. Parsing parity

From the outset, it is important to distinguish two different argumentative roles which PP has been made to play in support of HEC. In its first role, PP is a preparatory epistemological move to shift the burden of proof to the psychological internalist, challenging her to show why the biological boundary of the organism (or the brain) should matter for the demarcation of cognition. In its second role, PP is meant to provide a theoretically relevant criterion for individuating cognitive states. As we have said earlier, the criterion endorsed by Clark & Chalmers draws on the functionalist tenet that cognitive states are individuated in terms of the causal role which they play for the production of intelligent behavior. Let us refer to the first role of PP as its *epistemological* role, and to its second role as the *individuating* role. It is not exactly perspicuous to pack both argumentative roles into a single principle, since what it takes to succeed in each role is markedly different, as I will argue. Conversely, a failure to distinguish clearly between these two roles can easily lead to unwarranted criticisms of PP. Let me begin by circumscribing both argumentative roles in greater detail.

Elaborating its epistemological role, Clark (2005a, n. 3) has instructively labeled PP as a "veil of metabolic ignorance" in analogy to the "veil of ignorance," an argumentative device originally used by Rawls (2001) in his social contract theory of justice. Rawls invokes the veil of ignorance to define an impartial point of view – the "original position" – that we ought to adopt in our deliberations about the fundamental principles of justice. The point of Rawls' construction is to determine what kinds of principles people *would* jointly accept if they *were* negotiating as free and equal agents behind a presumed veil of ignorance, i.e., deprived of any information about their personal characteristics (e.g., their ethnicity, gender) and their particular social and historical circumstances. In order to ensure its neutrality, the veil of ignorance itself need not – in fact, should not, so as to avoid circularity – embody a substantive conception of justice. Similarly, the epistemological role of PP is merely to constrain what kind of information we can justifiably appeal to in a demarcation principle, without already presupposing a question-begging conception of cognition. In its first role role, PP is best understood as an egalitarian "fair-treatment" principle (Sprevak, 2009, p. 505) that advocates theoretical neutrality with respect to internal and external realizations of mental states. Understood as a purely meta-theoretical principle, we cannot expect PP to yield a substantive criterion for individuating cognitive states – a task left for the second role for which PP has at times been pressed into service.

Let us now consider the second, individuating role of PP. In order for the counterfactual comparison invoked by PP to make sense, it must in principle be possible that there can be extended and non-extended realizations of the very *same* cognitive states. Such a possibility is guaranteed by the functionalist thesis that "what makes something a mental state (of a particular type) does not depend on its internal constitution, but rather on the way it functions, or the role it plays, in the system of which it is a part" (Levin, 2010). The functional role of a mental state of type M is typically characterized by the psychologically relevant causal relations that M bears to sensory inputs, behavioral outputs, and other mental states. For instance, if we consider pain as an example of an organism-bound mental state, we might characterize the typical functional role of pain as follows:

> "Pain is the state that tends to *be caused by* bodily injury, and *causes* the belief that something is wrong with the body and the desire to be out of that state; it also tends to *cause* anxiety, and, in the absence of any stronger, conflicting desires, wincing and moaning" (Levin, 2010).

For a functionalist, the realization of any state with the same pattern of typical causes and effects counts as a pain-state; regardless of whether it is realized by the human nociceptive system, by an artificial creature with a very different physiological makeup, or by hybrid ensembles which are partly organic and partly non-organic. Within the philosophy of mind, the putative substrate-neutrality of mental states is known as *multiple realizability*. Since PP evidently requires multiple realizability, and functionalism serves multiple realizability on a silver platter, it makes to sense for Clark and Chalmers to wed the HEC and functionalism. The resulting position has recently been dubbed *extended functionalism* (Clark, 2008; Kiverstein & Clark, 2009). Against Shapiro's (2004) claim that the embodiment of cognition requires a rejection of multiple realizability, Clark (2008, pp. 203f) has recently reiterated his commitment to extended functionalism, emphasizing that

> "What matters [for the HEC-theorist] is just the full suite of encodings and operations made available by some combination of neural, gross bodily, and worldly op-

> portunities. Creatures with radically different bodies, brains, and worlds from us might thus contrive to use their varying resources to implement many of the very same cognitive and information-processing routines."

PP is prima facie compatible with two different strains of functionalism, depending on how we take the relevant functional roles of mental states to be defined (Block, 1980). According to analytic functionalism, the functional role of cognitive state types follows from a careful analysis of our corresponding commonsensical concepts. In contrast, psychofunctionalists hold that we ought to extrapolate the relevant functional roles from mentalistic explanations delivered by empirical psychology. It is conceivable that different psychological predicates require the use of different methods (cf. Clark, 2008, pp. 88-89). For instance, in order to determine whether some external portion of the environment (e.g., recordings in Otto's notebook) constitutes a dispositional belief, it may be indispensable to attend to commonsensical belief-desire explanations. With respect to cognitive phenomena at the periphery of our commonsensical understanding of the mind (e.g., motor control), our interpretation of PP may be more firmly guided by an appeal to scientific psychology. The suggested context-sensitivity of the individuating role of PP, as PP is applied to specific psychological predicates, is perfectly consistent with our "big tent" approach to the HEC (Chapter 1.4).

To underscore the difference between the two suggested roles of PP, let us draw out the implications of what it means to reject PP in each case. If somebody rejects PP as a fair treatment-principle, she asserts that even if some part of the world functions (in the context of a given task) as a process which, were it done in the head, we would recognize as cognitive, *it is nevertheless false* that this part of the world is therefore part of the cognitive process. Rejecting parity implies a substantial departure from a functionalist theory of the mental. It is thus incumbent on the critic who rejects PP to say what extra ingredients of mental states are essential to their possession, and what makes the brain uniquely apt to host such properties. Criticisms of this sort have been leveled against HEC from several directions.

For instance, Gertler (2007) has argued that only conscious, introspectible mental states should count as part of one's mind, because of their special role in the origination of genuine intentional actions. More generally, philosophers of a (neo-) Cartesian ilk have argued that consciousness sifts the mental wheat from the physical chaff, and that functional characterizations of mental states fail to capture the qualitative, self-intimating character of experiential mental states (Nagel, 1974; Chalmers, 1996). Some naturalist philosophers with dualist inclinations have speculated that it may be a primitive biological property of the brain to give rise to and sustain conscious mental states (Searle, 1984, 1992;

O'Connor, 2000). Consequently, it would at least be a matter of nomological necessity that mental states are only causally produced by brain states. In response to this criticism, Clark (2008, pp. 161-62; 2009; 2010, p. 56) has conceded that it remains a consistent (though to his mind unattractive) option to "shrink" the realm of the mental to the occurrence of conscious mental states. At the very least, this way of rejecting PP would be a violation of EXTENDED FAIRNESS (Chapter 1.5). It would imply that the kind of "tacit" mental representations which figure prominently in orthodox cognitive-scientific explanations are not part of one's mind, because they are not (even in principle) accessible to consciousness either (Stich, 1978; Marr, 1982; Chomsky, 1986). From a very different perspective, Wheeler (2010b) argues for an incompatibility between certain strands of enactivism (e.g., Di Paolo, 2005) and extended functionalism if the notion of *enaction* is intimately tied to the capacity of organism-bound living systems for autopoiesis and sense-making (see also Kiverstein and Clark, 2009). Like the claim that purely functional characterizations of mental states cannot account for the phenomenal aspects of consciousness, radical enactivism also appears to entail a full-blown rejection of minimal functionalism (in this context, see also Shapiro, 2004). This puts both positions outside the scope of objections that are consistent with our maxim of EXTENDED FAIRNESS (Chapter 1.5).

Does this leave any room for the critic who accepts PP as a "fair treatment" principle and also endorses a broadly functionalist view of the mental, yet still wants to resist the argument for the HEC? As opposed to what we have called the *rejection* of PP, let us call this option an attempt to *block* the implications of PP further downstream. The most promising way of blocking PP is to argue that external processes are in fact functionally quite *unlike* their corresponding counterparts inside the brain; hence they are not cognitive. To block PP in this manner, it is incumbent on the critic to articulate in greater detail the specific functional roles which they take to be necessary and sufficient for the realization of certain mental states. Critics who have shouldered this burden have made the case that the HEC succumbs to the following dilemma (e.g., Rupert, 2004, pp. 22-39; 2009, pp. 96-99). If one requires a sufficiently fine-grained specification of the relevant functional roles, it is unlikely to be matched by many extended processes. On the other hand, suppose the HEC-theorist leans on a more coarse-grained functional profile capable of subsuming both internal and extended realizers. Then the critic retorts that such generic cognitive kinds do not support any interesting inductive generalizations, and thus can't do any serious causal-explanatory work in psychology.

Let us now take a closer look at the current debate in light of our suggested reconstruction.

2.3. Parity, similarity, and functional equivalence

Several critics have objected to Clark and Chalmers' argument from PP on the grounds that their imagined cases of extended cognition are in many ways relevantly *unlike* their familiar cognitive counterparts inside the brain. For instance, Adams and Aizawa attribute to Clark and Chalmers a view of PP according to which "the active causal processes that extend into the environment are *just like the ones found in intracranial cognition*" (2001, p. 56). In order to disprove that claim, they go to great lengths to show that intracranial and transcranial processes are by no means functionally equivalent. As one of their examples, they point out that biological memory "displays a number of what appear to be law-like regularities, including primacy effects, recency effects, chunking effects and others" (ibid., p. 61), none of which are displayed by the gerrymandered set of external "memory" devices that include calendars, CD-ROMs, and Otto's notebook. Also, Otto's act of writing down notes requires the use of his motor system (e.g., when he uses his fingers to flip through the pages), his visual system (e.g., when he reads the entries), and a variety of beliefs about the physical and functional makeup of the notebook. Since the exact causal profile by which Otto encodes and retrieves information in his notebook is quite different from Inga's access to her biological memories, they conclude that it would be misleading to lump the two together as cognitive states (or processes) of the same kind.

Rupert (2004, pp. 409-415) similarly reviews a wide range of psychological data on human memory performance to "quell[s] any temptation to argue for HEC from brute analogy (namely, extended cognitive states are like wholly internal ones; therefore, they are of the same explanatory cognitive kind; therefore there are extended cognitive states)" (ibid., p. 407). As a case in point (ibid, p. 413f), when human subjects are asked to learn information anew which they had previously memorized in a different setting, they are subject to *negative transfer*. For instance, if a subject first learns to associate pair-wise items A-B, it takes her significantly longer to learn new associations of the form A-C, because what was previously learned in one context seems to hinder or delay learning in a new context. However, it is not any harder for Otto to write down and recall a novel pair A-C than it was earlier for him to write down and recall A-B. For the HEC-theorist, this raises the problem of explaining why "particular learning curves, interference patterns, and so forth, are unavoidable when we rely on internal storage, while entirely optional, in fact difficult to locate or produce, in the case of external storage" (ibid., p. 418).[15]

Other critics have emphasized the active, i.e., integrative and reconstructive role of biological memory. Compared to our biological memories, so the critics argue, the inscriptions in Otto's notebook are informationally isolated from the rest of Otto's mental states (cf. O'Brien, 1998; Dartnall, 2006). Weiskopf (2008) forcefully illustrates this point with the following example. Let us suppose that when Inga hears that MoMA has recently been relocated to Brooklyn, she automatically revises her prior belief that MoMA is on 53rd Street. This causes a series of further changes in her doxastic system, such as discarding the belief that MoMA is only a stone's throw away from her apartment on 52nd Street. All of these inferential connections are part of the functional role by which functionalists individuate memory states. The inscriptions in Otto's notebook, however, are not subject to any such automatically performed and actively propagated revisions. Otto's notebook is filled up with inferentially isolated bits and pieces of information. Because of the lack of informational integration, treating the notebook as part of Otto's extended mind would mean that one would have to attribute a large set of wildly incoherent beliefs and desires to Otto-the-extended-cognizer. This would imply that Otto-the-extended-cognizer exhibits a degree of irrationality which is very different from normal people. Consequently, since the inscriptions in Otto's notebook are not functionally equivalent to the states of Otto's biological memory, we should not treat them as instances of the same cognitive kind.

The common goal of these objections is to meet the HEC-theorist on its own functionalist turf, but use the functional disparity between inner and outer resources to block the inference from PP. We can spell out this criticism in the form of a counter-argument:

1. For two states/processes to be of the same cognitive kind, they must be functionally equivalent.
2. For two states to be functionally equivalent, they must share their fine-grained functional role.
3. Internal and external "memory" processes do not share their fine-grained functional role.

Therefore, internal and external "memory" processes are not of the same cognitive kind.

15 A similar point is made by Adams and Aizawa (2001, pp. 58-61).

To deflect this criticism, proponents of the HEC have tried to undermine the conjunction of Premises (2) and (3) by describing carefully imagined cases in which the supposed functional differences between inner and outer resources play an increasingly marginal role. For instance, in response to Weiskopf's "active integration"-argument, Clark (2005a, 2010) has suggested to think of Otto along the lines of Leonard, the main character in Christopher Nolan's captivating movie *Memento* (2001). Leonard's ability to store new long-term memories is severely hampered by an untreatable form of anterograde amnesia. Fueled by an all-consuming desire to track down the man who raped and murdered his wife, Leonard tries to compensate for his mental handicap by compiling a huge inventory of external stand-ins – notes, photographs, file folders, and charts. He even gets body tattoos that engrave what he believes are the most important pieces of the puzzling murder case. Leonard's greatest challenge is to graft the newly obtained bits of evidence into his growing storage of information – trying frantically to edit, alter, and remove outdated memory entries as he tries to uncover the truth about his wife's murder.

Now imagine another creature – let's call him Otto* – whose *internal* memory management resembles the epistemic procedures employed by Leonard, but executed to perfection. Viewed from the standpoint of input-output equivalence, Otto* acts just like Inga, a person with a normally functioning biological memory. The only difference between the two is the internal mechanism by which the desired integration of hitherto segregated bits of memories is achieved. Unlike Inga, Otto* does not immediately update the inferential connections between his memories when he acquires new information. Instead, his brain does it just before outdated or otherwise incoherent information would be accessed by higher-level cognitive processes (e.g., an executive decision to initiate a certain intentional action). As Clark points out, Otto*'s procedure bears a certain resemblance to what is known in computer science as "dynamical" or "just-in-time" compilation. Now let us suppose further that Otto efficiently performs all the relevant procedures "outwardly" that Otto*'s brain is executing "inwardly." If somebody asks Otto whether he wants to have a cup of coffee at MoMA, he never answers without scanning the relevant portions of his notebook. If he detects an inconsistency between an older and a more recent entry, he resolves it by deleting the old entry before taking on the invitation, and updates and revises other beliefs accordingly. As long as Otto can make his procedure work at run-time, his actions need not be any less rational than Inga's.

In addition, Clark has pointed out that there are idiosyncratic cognitive domains in which our biological memory is unusually fragmented and thus would not meet Weiskopf's "active-integration" requirement either (Clark, 2005a, p. 7; 2010, p. 53). For instance, persons with an eidetic memory can exercise their extraordinary abilities only if the stimuli to be recalled are encoded in a specific representational format (e.g., visual or auditory). Another example is our capacity to rote-learn large chunks of written text. Rote-learned sentences exhibit recall characteristics which are quite distinct from regular text comprehension (e.g., we often have to regurgitate the first five lines when prompted to recite the sixth line of a Latin poem). None of these internal states exhibits the active kind of informational integration which may be typical of many biological memory phenomena, and yet we do not hesitate in treating these internal states as part of our memories.

The upshot of Clark's response is that when we treat two mental states as functionally equivalent, we do not always care about all the gory details concerning the ways in which they are processed. We seem to have a license to ignore fine-grained functional differences if we allow, as functionalists do, that different mechanisms can on different occasions realize one and the same type of cognitive processes. We effectively "black-box" those mechanistic differences when we assert, in the context of explaining their shared disposition to walk towards 53rd Street, that Otto and Inga share the belief that MoMA is on 53rd Street. The suggested flexibility in interpretation is surely an important part of what makes the intentional stance so eminently useful (Dennett, 1987).

As Sprevak (2010, pp. 510-514) observes, it is an implicit feature (or a bug, according to some – including Sprevak himself) of functionalism that it contains a "parameter" which controls how finely or coarsely functional roles are specified. For instance, we omitted from our above specification of the typical pain-role that the loss of a person's significant other is a typical cause of social pain in human beings, a type of mental state that is known to recruit some of the same neural mechanisms underlying physical pain (Robbins, 2008). Similarly, rather than enumerate all typical causes of physical pain such as cuts, burns, and gastric upsets one by one, we abstracted away such details to form the more generic kind *bodily injury*. Our reason for doing so is that not all idiosyncratic features of human pain seem to be essential for ascribing pain to an organism. If we set the grain parameter too fine, this would lead to an undesirable loss of generality in the scope of our folk-psychological explanations. Consequently, at least as long as we are concerned with folk-psychological predicates, it is implausible to tie the benchmark for PP too close to any specific details of brain-bound human cognition. Hence we do not have to require perfect similarity between intracranial and transcranial processes for establishing their functional equivalence as cognitive kinds.

Does this inevitably lead to a deadlock between extended functionalism and its critics, given that there does not seem to be a unique way of resolving the question of how we ought to set the functional grain parameter? Rowlands

(2008) has argued that the looming stalemate threatens to paralyze the entire debate over the HEC: "If Rupert's arguments against the extended mind are question-begging because they presuppose a chauvinistic form of functionalism, it is difficult to see why arguments for the extended mind are not question-begging given their predication on a liberal form of functionalism" (ibid, pp. 6-7; cited after Wheeler 2010a, p. 255). According to Rowlands, the indeterminacy of the choice between chauvinistic and liberal specifications of functional roles is an issue that has plagued functionalism since its inception, hence there is not much hope that extended functionalism can be conclusively proven or refuted.

However, contrary to Rowlands, and in basic agreement with Wheeler (ibid.), it would be premature to conclude that "anything goes" in our settings of the grain parameter. In particular, I think there are independently motivated considerations that speak against setting the relevant level of functional grain too closely to the fine details of the human inner. Dennett (1993b) once argued that if we were to discover that left-handers, or women, used virtual brain machines for some cognitive tasks for which other people used real brain machines, this would not undermine our folk-attributions of mentality to both kinds of people, even with regard to the competence in question. Underlying this intuition is not just Dennett's cavalier attitude towards folk-psychology, but a deeper methodological point that applies equally to scientific versions of psychofunctionalism (see also Wheeler 2010a, pp. 256f.). Let me try to flesh out this point with a concrete example.

Cognitive psychologists are generally interested in comparing the performance of cognitive capacities under different conditions. For instance, there are a large number of studies which attempt to measure the impact of "belief bias" on our deductive reasoning competence. In these studies, the believability of a given conclusion is systematically manipulated alongside the validity of the argument in which it occurs (Evans, Newstead, & Byrne, 1993). The results of these studies reliably show, among other things, that people are far more inclined to reject invalid+unbelievable arguments than invalid+believable ones. Now, is the observed interaction between belief bias and logical validity an essential feature of human deductive reasoning? I think not. If this were the case, we could no longer truly say of students who finally rid themselves of belief bias – perhaps because they took an introduction to critical reasoning – that they still engage in deductive reasoning. This is a fairly unintuitive implication that belies our educational efforts to improve students' deductive reasoning skills. Surely the discovery of abnormal student populations who are completely immune to belief bias would prompt an avalanche of further investigations. Cognitive psychologists would be licking their chops, and run the whole gamut of experiments to figure out questions like: *How come these subjects systematically ignore cues about the believability of information? What makes these deductive reasoners tick? Do they perhaps utilize an altogether different neural mechanism? Why did they develop such a mechanism?* But for any such comparison between normal and abnormal subjects to be informative, cognitive psychologists must presuppose that subjects from both populations do in some broad but significant sense engage in deductive reasoning. Otherwise, why bother running these studies, if they only compare apples and oranges? The same basic point underlies the logic of similar comparative studies of cognition, such as comparing the performances of normal and impaired subjects, of novices and experts, of children and adults, or of humans and animals with respect to some cognitive capacity X. They all require that there is some broadly defined equivalence class of cognitive processes or capacities relative to which the two different populations of subjects can be compared. This is precisely Dennett's point all over again.

In sum, it would be premature to declare that the debate over extended functionalism suffers from a lethal kind of indeterminacy. Granted, it is implausible to think that there must be a single correct setting of the functional grain parameter that equally applies to all types of cognitive states (or processes); but we have already emphasized that extended functionalism ought to be relativized to specific psychological predicates, applied to particular situations. What I have tried to show is that there are many perfectly legitimate contexts of explanations in which we do not demand fine-grained functional similarity for sameness of cognitive states, because it would simply slice the cognitive pie too thin.

2.4. Parity, parsimony, and folk psychology

A related criticism of PP concerns the role of perception and action in demarcating the bounds of cognition. Traditional accounts of cognition often depict the mind as a sort of "cognitive sandwich" (Hurley, 1998b) wedged between perception and action.[16] From the acceptance of this picture it is only a short step to argue that the cognitive sandwich must be lodged inside the brain – skin and in – between sensory and motor processes. Thus anything which precedes sensation, and anything that follows action, cannot be truly cognitive. The relationship of cognition to perception and action thus serve to highlight another potential functional asymmetry between the ways in which Inga and Otto access their respec-

16 This picture of the relationship between sensing, thinking, and acting has been challenged by theories of embodied and extended perception, including those found in Gibson (1979), Varela, Thompson, & Rosch (1991), Thelen & Smith (1994), Churchland, Ramachandran, & Sejnowski (1994), Hurley (1998b), O'Regan & Noë (2001), Noë (2004). However, those criticisms are mostly orthogonal to the issue that I am about to discuss.

tive memory resources. When Otto interacts with his notebook, his actions involve the use of his motor system (when he flips through the pages) and his visual system (when he reads the entry), and depend on a variety of beliefs about the makeup, trustworthiness, and availability of the notebook. Otto's notebook-manipulations involve perception and action in ways that are simply not required for Inga to access her internal memories. Similarly, in the game of Tetris, the epistemic actions performed by Player 2 involve a host of additional perceptual, motor, but also inner psychological processes that are not required for Player 1 (Chapter 2.1). For instance, the muscle activity which is necessary to push the button on the console requires the activation of motor processing systems, and it involves an executive decision – made by the brain of the cognitive agent – as to which of the two available methods of rotation to use (cf. Adams & Aizawa, 2008, p. 119). Because of these asymmetries of access, one might object to the claim that Otto's external manipulations are functionally equivalent to human cognitive processes.

Anticipating this objection, Clark & Chalmers (1998, p. 16) argue that Otto's access to his notebook should not be properly viewed as a case of *perception*, but is functionally more akin to *introspection*. In similar vein, Chalmers (2008, p. ix) suggests that there can be instances of inner perception (e.g., inspecting a mental note) and inner mental action (e.g., making a mental note), and Otto's dealings with his notebook are functionally equivalent to those. More generally, extended functionalists can argue that it would be question-begging to identify the relevant "inputs" to cognition with sensory stimuli impinging on the surface of the organism, and the relevant "outputs" with episodes of motor behavior. Such identification would unduly tie the boundaries of a cognitive system to the sensorimotor interface of the organic body (cf. Wheeler, 2010a, p. 249). It is precisely the claim of extended functionalism that some episodes of sensory and motor processing can be properly subsumed under a psychological description. Hence they should be *functionally* defined just like every other psychological predicate.

At least for the case of intentional mental states, the suggested functional redescription is plausible to the extent that folk psychology ignores the difference between sensory perception and inner perception, or between motor action and inner action. However, it is quite clear that folk psychology does at least sometimes mark such a distinction. For instance, Rupert (2004, pp. 91-94) has argued that commonsense psychology rebels against the HEC, claiming (inter alia) that our folk conception of belief precludes its being *seen* by its possessor. In response, the HEC-theorist might order a healthy dose of revisionism (cf. Chalmers, 2008, p. x). The standard move here is to insist that Otto's notebook entries share enough of the central features that ordinary beliefs play in folk-psychological explanations. Consequently, we *ought* to treat them as beliefs, regardless of whether all the folk concur or not.

There is a way of turning this somewhat indecisive consideration into a more serious objection to the HEC, one that was also anticipated by Clark & Chalmers in their original paper (1998, pp. 13f). Because the objection is not only that Otto's access to his notebook is invariably mediated by physical actions and sensory perceptions, but also that there is a perfectly good psychological reason for those sensorimotor interactions – namely Otto's non-extended state of *ignorance* about the location of MoMA. Hence instead of ascribing to Otto the extended belief that MoMA is on 53rd Street, there is an equally good, but ontologically more conservative folk-psychological explanation of Otto's behavior. Because Otto desires to go to MoMA, but does not know its location, he forms the non-extended belief that the address of MoMA is probably written down in his notebook (= Step 1). This leads him to reach for his notebook, and attentively flip through its pages. When he finds the relevant entry, he forms the non-extended belief that MoMA is located on 53rd Street (= Step 2). So if we individuate mental states by the role which is accorded to them in commonsense psychology, the availability of an "Otto two-step" explanation (Clark, 2010, p. 46) would seem to wield evidence against the HEC.

Against the Otto-two-step, Clark and Chalmers (loc. cit.) argue that a similar interpretation is also available with respect to Inga's access of her biological memories, but it would (implausibly) imply that those memories are not literally speaking part of Inga's mind either. Rather than saying that Inga remembers that MoMA is on 53rd Street, one might equally say that Inga believes that her biological long-term memory contains information about the location of MoMA, and only upon retrieving the relevant item into her working memory does she come to believe that MoMA is located on 53rd Street. In fact, there is evidence that people tacitly invoke so-called meta-memories – i.e., memories about the content, location, and reliability of their own memories – when they try to remember things. An important early stage of memory retrieval is to assess one's "feeling of knowing" (Hart, 1967), which has been shown to increase the speed and success rate of retrieving information from personal memory (Flavell & Wellman, 1977; Nelson & Narens, 1990). But for the global purpose of explaining why Inga walked up north to 53rd Street (rather than downtown), citing Inga's consultation of her own meta-memories would pointlessly add psychological complexity. An Inga-two-step would not make Inga's overall behavior any more intelligible to a third party than just citing the content of her ordinary (first-order) memories. In the suggested context of explanation, considerations of parsimony favor an Inga-one-step analysis. However, suppose we want to know why Inga hesitated in her decision to walk north, and we are told because she wasn't sure whether she accurately remembered the location of MoMA. In this case, the Inga-two-step does add contextually relevant psychological complexity to our story, and thus ought to be preferred (cf. Chalmers, 2008, pp. x-xii). And what's true of Inga is certainly true of Otto. Suppose we ask: why did Otto reach for his notebook? To explain this episode of local-scale behavior, a perfectly good answer is to say that he *didn't* remember the location of MoMA, but believed that it was written down in his notebook. But if we simply aim to account for Otto's large-scale behavior, it is perfectly natural to leave out Otto's dealings with his notebook as psychologically redundant, and consider the notebook as part of his extended memory.

In conclusion, the Otto-2-step does not show anything beyond the fairly uncontroversial claim that our attribution of mental states, according to "commonsensical" functionalism, is sensitive to the context of psychological explanation. A change in context often triggers a corresponding switch in our setting of the "functional-grain parameter," which can significantly alter the semantic extension of the relevant psychological predicate. However, this does not disprove the HEC, as long as there are enough central contexts in which extended representational states function very much like beliefs in ordinary psychological explanations.

2.5. Parity, similarity, and functional integration

In the previous sections, I have defended the idea that a fairly coarse-grained functional similarity between neural and bio-external resources can warrant a unified cognitive treatment. But some proponents of the HEC have purported that there is an altogether different understanding of the relationship between inner and outer resources which is completely devoid of this questionable presupposition, and thus affords a logically independent path to the claim that cognition is extended (cf. Menary, 2006, 2007; Sutton, 2010).

To develop this alternative line of reasoning, let us revisit the case of memory. HEC-theorists have certainly not been oblivious to the fact that inner and outer memory resources typically do not share many of their attributes, and often serve very different functions in our cognitive economy. A nice example is Merlin Donald's (1991) theory of "exograms" – i.e., external representational artifacts such as body tattoos, grave decorations, sculptures, hieroglyphics, maps, graphs, and writing systems. Donald has argued that the spread of exograms was responsible for the last major transition in the evolution of human cognition, citing the critical role which external methods of memory storage, coordination, and retrieval have played in sculpting the cognitive architecture of our modern minds.[17] Clearly, Donald's evolutionary hypothesis makes sense only on the assumption that the typical functional role of exograms, and the procedures we have developed for using them are crucially different from how we form and manipulate "engrams" in our biological memory (Table 1).

Table 1 (based on Donald, 1991, p. 315). A comparison between engrams and exograms.

ENGRAMS	EXOGRAMS
Internal memory record	External memory record
Fixed physical medium	Virtually unlimited media
Constrained format	Unconstrained and reformattable
Impermanent	May be permanent
Large but limited capacity	Virtually unlimited
Limited size of single entries	Virtually unlimited
Not easily refined	Unlimited iterative refinement
Retrieval paths constrained	Retrieval paths unconstrained
Limited perceptual access in audition, virtually none in vision	Unlimited perceptual access, especially in vision; spatial structure useful as an organizational device

Setting aside possible borderline cases such as Otto and his notebook, doesn't the centrality of this fairly coarse-grained contrast undermine the version of extended functionalism that we have defended? To forestall an objection of this kind, Richard Menary (2006) has warned that the theoretical weight which PP puts on functional similarity is seriously flawed, and greatly distracts from what he takes to be the core insight of the extended mind thesis. According to Menary, the central goal of the extended mind thesis is to foster the project of "[...] 'cognitive integration,' which is the view that internal and external vehicles and processes are integrated into a whole" (ibid, p. 329). The main take-home message of Menary's emphasis on cognitive integration is that an extended mind is always a deeply *hybrid* mind – insofar as the functionality of internal and external resources that cognitive agents actively incorporate into their problem-solving routines are typically *dissimilar*, but complement one another (p. 330). This implies that our ability to perform cognitive tasks by manipulating environmental structures is considerably *different* from our untutored ability to tackle the same task with our indigenous, biologically-based cognitive resources. We tend to forget how tedious the enculturation process is for creatures with biological brains like ours; how much effort it takes to teach somebody how to read and write, or to juggle long-term multiplication problems with the

17 Rowlands (1999, p. 133ff) presents a detailed argument for extended memory that refers to Donald (1991).

partial-products method. What makes our most prominent cognitive artifacts such as pocket calculators, Arabic numerals, and word-processors such indispensable resources is that they support operations at which our "naked" brains are naturally bad (Norman, 1988, 1993; Clark, 2003; cf. Chapter 1.2.3). Teachers have to work hard to instill into their students such erudite practices of environmental manipulation, together with the cognitive norms governing those practices. Once we rest our focus on the ways in which cognitive *integration* is achieved in our "mashup" minds (Clark, 2008, p. 218), the issue of *similarity* between the inner and the outer becomes a red herring.

It is for this reason that Clark and Chalmers' presentation of their thought experiment, which emphasizes the functional analogies between Otto-and-his-notebook and Inga's biological memory, may have done a slight disservice to their own cause. Their example inadvertently shifts our attention to the limited functionality of the notebook as a poor *substitute* of biological memory, e.g., for its lack of informational integration (Weiskopf, 2008). Notice, for instance, that Otto is said to rely on his notebook primarily because he needs it to *compensate* for his partial loss of biological memory. If our goal is to replace a lost or malfunctioning part of a pre-existing system, our search for an equivalent substitute is driven by the requirement that the new part should be able to perform the same (or at least very similar) key functions as the old part. This expectation may have primed the questionable assumption that PP is based on the requirement of fine-grained functional similarity.

Other proponents of the HEC have similarly honed in on cognitive integration as the cornerstone of the extended mind. In his analysis of memory as an extended cognitive phenomenon, Rob Wilson (2004) emphasizes that "remembering, on this view, involves exploiting internal, bodily, and environmental resources in order to produce some sort of action, often social in nature" (p. 191). In order to signal his departure from Clark & Chalmers' argument from PP, Wilson (2010, p. 173) suggests that we ought to distinguish between "cyborg fantasy" and "active cognition" arguments for HEC. According to Wilson, the theoretical focus of cyborg fantasy arguments lies on the technological amplification of our biologically predisposed cognitive capacities by external artifacts such as Otto's notebook, as well as either imaginary or soon-to-be-realized gadgets such as neural implants and tele-robotic devices (Clark, 2003). The family of active cognition arguments, on the other hand, is meant to highlight the critical role which the proper integration of biological and *complementary* environmental resources plays for the exercise of specific cognitive capacities (e.g., in perception, memory, problem-solving, cognitive development). Contrary to cyborg fantasies, active cognition arguments are based closely on empirical studies of cognition in its natural environment, with the goal of showing that much of human cognition as we currently know it is extended. According to this classification, the arguments found in Clark (1997, 2008), Hutchins (1995a), Haugeland (1998), Hurley, (1998b), Rowlands (1999, 2006), Wilson (1995, 2004, 2010), Noë (2004, 2009), Sutton (2006, 2010), Menary (2006, 2007), and Theiner (2007) fall into the category of active cognition arguments.

Let us spell out the argument from cognitive integration in greater detail, in order to bring out how the concomitant shift in emphasis concerning the role of PP. My reconstruction of the argument is based closely on Wilson's own "argument from meaning-making," which he subsumes under the category of active cognition arguments:[18]

1. The exercise of a cognitive capacity M by some cognitive system S involves the sequestering and integration of internal and external resources.
2. Internal resources are part of the structure of S in virtue of their functional contributions to the exercise of M by S.
3. External resources must be functionally integrated with internal resources for the exercise of M either insofar as they play similar functional roles, or because they complement, compensate for, or enhance these roles.

Therefore, external resources, like internal resources, are equally part of the structure of S which as a whole is causally responsible for the exercise of M.

What role, if any, does PP play in the argument from cognitive integration? In his own formulation of the argument, Wilson does not include PP as a separate premise, although his subsequent gloss indicates that he is well aware of the fact that the suggested inference relies on it: "This [i.e., the conclusion] follows from a parity assumption, namely, that things that have the same relevant properties vis-à-vis some cognitive process should have the same status in cognition as one another" (Wilson, 2010a, p. 175). I take Wilson's gloss to mean that internal and external resources are said to be equally cognitive because they both make causally distinctive, functionally indispensable contributions to the exercise of the cognitive capacity in question. As Premise (3) states, this does not imply that their respective contributions ought to be similar – they can also stand in a complementary, compensatory, or otherwise transformative relationship to each other.

---

18 Wilson (ibid.) frames his argument in terms of intentionality, but we can easily substitute for any other other cognitive capacity provided the goal is to show that "it is only the whole functioning, beyond-the-head causal system that constitutes the matter in motion that realizes the exercise of that capacity" (Wilson, 2010a, p. 173).

However, this can't be the whole story. While the argument from integration certainly highlights the productive interplay between inner and outer resources from an interesting new angle, it seems that PP continues to play a more critical role than Wilson is willing to admit here. This is because the critic can easily force the issue, and demand a further reason why we should consider the external resources as part of a single, transcranially extended *cognitive* process. To this objection, the integrationist will most likely cite Premise (1), claiming that what makes the external parts of the transcranial process as cognitive as its internal parts is the fact that they all make *some* distinctive, functionally indispensable causal contribution to the exercise of a cognitive capacity. But this only postpones the problem, because the critic can now consistently demand to know what marks that capacity as a *cognitive* capacity in the first place. And I don't see how extended functionalists can respond to that challenge without appealing to some version of PP in order to individuate the relevant capacities. For instance, a satisfactory answer would be that some capacity M is cognitive because it is relevantly similar to other, intracranial instances of its kind (e.g., memory, decision-making, motor control) that we have independent reasons to consider as cognitive. But this, of course, means that PP has quietly been let back in through the back door.

2.6. Clark's parity versus Turing's parity

Clark has recently clarified his preferred interpretation of the HEC, saying that it is meant to support

> "the claim [...] that taken as a single, integrated system, Otto-and-the-notebook exhibit enough of the central features and dynamics of a normal agent having (amongst others) the dispositional belief that MoMA is on 53rd Street to warrant treating him as such" (Clark, 2005a, p. 7).

In an effort to unpack his "commonsense-functionalist" understanding of PP a bit further, Clark (2010) has proffered a quasi-operational criterion according to which extended mental states of a cognitive agent

> "should guide behavior, reason, and choice in ways that would not cause constant misunderstandings and upsets if the agent were somehow able to join with, or communicate with, a human community" (Clark, 2010, p. 62).

Rupert (2009, p. 32) has taken Clark's increasingly liberal version of PP to argue, against the HEC, that it would succeed at demarcating cognitive from non-cognitive states only at the price of turning the notion of *cognition* into a response-dependent concept. Roughly speaking, response-dependent concepts are concepts whose truthful application is grounded not in objective matters of fact, but in our human disposition to apply them to certain things in certain ways under certain (perhaps idealized) conditions (Pettit, 1991; Casati & Tappolet, 1998). For instance, assuming for the sake of the argument that *being cool* is a response-dependent concept, the fact that some person *S* is cool would be determined by the disposition of a sufficient number of people to judge that *S* is cool. Similarly, if the point of PP is simply to divide the universe into things that have the property to elicit the judgment *cognitive* from enough suitably qualified human observers and things that don't, the concept of *cognition* would be response-dependent. If Rupert is correct, such an implication would have problematic consequences for the HEC for at least two reasons. First, the absence of any substantive, i.e., response-*in*dependent properties of scientific importance ("natural kinds") that could justify our assent to PP would seriously diminish the prospects for a science of the mind. Hence if HEC proclaims to spearhead the embodied and embedded revolution in cognitive science, the suggested implication would quickly put an end to those fantasies. Second, the apparent lack of consensus on our judgments about PP – perhaps even among suitably qualified observers – would mean that PP cannot be trusted to play a theoretically central, decisive role in arguments for the HEC. If anything, PP might be used to instigate further experimental investigations of people's intuitions about what's cognitive (Knobe & Prinz, 2008).

In this section, I want to alleviate this potential worry by contrasting Clark's understanding of PP with the behavior-based criterion of cognition ("intelligence") that is employed in the Turing Test (Turing, 1950). Considering the vexed question of whether machines can think, Turing famously refused to give a general definition of the terms "think" and "machine," and instead suggested that we "replace[s] the question by another, which is closely related to it and is expressed in relatively unambiguous words" (ibid., p. 433). He then proposed a so-called "imitation game" in which a human judge is summoned to carry out a potentially open-ended online chat with a human being in one room, and a machine in another room. If the judge cannot reliably tell who's who over a series of trials, the machine is said to have passed Turing's test, and should thus be considered as intelligent (read: as able to *think*). Turing was convinced that his test has "the advantage of drawing a fairly sharp line between the physical and the intellectual capacities of a man" (ibid., p. 434). The suggested criterion of cognition can be expressed as a kind of parity principle:

TURING-PARITY (TP): If some system *S* is able to pass Turing's test which, *were it passed by a human being*, the judge would accept as an intelligent cognizer, then *S is* an intelligent cognizer.

In what ways, if any, does Clark's conception of PP differ from Turing's conception of TP? We can immediately check off a few obvious, but fairly superficial differences. First, the purpose of PP is not to compare the cognitive capacities of man and machine, but of an organism-bound cognizer (Inga) and an extended, hybrid cognizer (Otto-and-his-notebook). Second, whereas TP relies on a single person to make the decisive call, the quite from Clark suggests that PP ought to be judged against the backdrop of community-wide standards. But putting these differences aside, PP and TP certainly have some interesting features in common. First, neither TP nor PP aims to provide a general definition, or a conceptual analysis of the meaning of cognition. Nor do they attempt to give a substantive empirical theory of the central, distinctive properties of cognition – the putative "mark of the mental" (Adams & Aizawa, 2008). In order to pass Turing's test, we do not have to be able to find some conceptually necessary and sufficient conditions that any sequence of question-response pairs ought to satisfy in order to count as a genuinely "meaningful" and "intelligent" conversation. Similarly, PP is not committed to any specific traits that Otto-the-extended cognizer must possess in order to share "enough of the central features and dynamics of a normal agent" (Clark, loc. cit.).

Moreover, Rupert's criticism of PP strikes a chord with those who think that the purely operational criterion of intelligence/cognition favored by TP has serious drawbacks. TP self-consciously relies on more or less intuitive judgments of human observers, but we know that humans are predictably biased in their judgments about cognition. For instance, humans tend to discriminate against digital computers because of their overly impressive, "machine-like" perfection in performing cognitive tasks that are particularly hard for us (e.g., complex numerical calculations). On the other hand, humans can reportedly be fooled to impute mentality to computer programs which do nothing but employ a few cheap tricks (Weizenbaum, 1966). This dilemma has received a lot of attention in the philosophical literature on the descriptive adequacy of the Turing Test as a general criterion of cognition/intelligence (Oppy & Dowe, 2011). For our comparison between TP and PP, it will be helpful to review several strands of criticism.

From a functionalist perspective, the original Turing Test has been criticized for its residual commitment to a form of behaviorism, insofar as the judge is forced to rely on purely behavioral criteria ("input-output equivalence") to assess the outcome of the test. In order to bring out and criticize this limitation, Block (1981) conceived a machine (a "Block-head") that would be able to pass Turing's test, but in a way that we would strongly hesitate to consider as intelligent. Block's argument is based on the assumption that for any finite amount of time (e.g., 1 hour) during which the Turing Test is administered, only a finite number of strings of sentences (consisting of question-answer pairs) can be typed by the judge. Accordingly, it should be possible in principle to store all sequences of strings that would constitute a "sensible" conversation (for some finite amount of time) in a giant look-up table, and store it in the machine. When Turing's test is administered, the machine waits for some string A to be entered, then searches its list of A-initial sensible strings, and randomly picks out one as its response B. Then the interrogator types in C, so the machine searches for strings which begin with ABC; and so forth. Block argues that a Block-head would not be intelligent, even though it could in principle pass Turing's test.

Block goes on to reject TP as an empirically adequate criterion of intelligence/cognition in favor of *psychologism*, a position which he defines as follows. Consider two systems which are exactly alike in their behavioral dispositions, yet differ in terms of the information-processing mechanisms that mediate between the sensory inputs and behavioral outputs. Then psychologism can be defined as the thesis that this purely internal difference is enough to justify the claim that one system is intelligent, while the other is not. Psychologism is thus incompatible with behaviorism, because psychologism cares about *how* things are done in cognition. On the other hand, as Block stresses, psychologism should not be confused with anthropocentric "chauvinism" – the claim that only a very specific kind of internal cognitive architecture – e.g., the one displayed by human beings – is necessary (and sufficient) for genuine intelligence. Along similar lines, Rey (2002) has argued (in response to Searle's (1980) "Chinese Room" argument) that the Turing Test is not a decisive test for the computational theory of mind. Turing's criterion unrightfully saddles computationalism with the "black-box" methodology that was characteristic of behaviorism. Once we place further theoretical and computational constraints on what a system must be able to do in order to understand the potentially infinite structure of natural language, we can successfully rule out the "conversation manual" model of linguistic understanding that is implicitly assumed by the Turing Test.

Highlighting a different aspect of cognition, Dennett (1987) has stressed the importance of processing speed (defined relative to the salient features of a system's environment). For a real cognizer, "speed [...] is 'of the essence' for intelligence. If you can't figure out the relevant portions of the changing environment fast enough to fend for yourself, you are not practically intelligent, however complex you are" (ibid, 326). Consequently, even if a Block-head were somehow capable of producing appropriate answers *in principle* – especially if we relax all time constraints – those answers would come much too slow in a real-world environment, because of the extremely inefficient organization of the giant look-up table. In the real world, slow thinkers are stupid, not intelligent – they may well be dead before they have churned out the correct answer. On the other hand, so Dennett, if a computational system can be sped up enough to connect the sensory transducers and motor effectors of its (robotic or organic) body in time, it would be truly cognitive – regardless of its material substrate (p. 332). In somewhat similar vein, Harnad (1990, 1991) has criticized the self-imposed limitation of the Turing Test to focus on the narrow bandwidth of intellectual capacities which easily lend themselves to verbalization. He argues that the intelligent use of symbol-based intellectual capacities requires the functional integration of behavior-based capacities that would require the control of a real body. For Harnad, cognition must be perceptually grounded in sensory and motor capacities that causally connect a computational system with the world.

Interestingly enough, the Turing Test has also been criticized from an almost opposite perspective, namely for its undue anthropocentricity – its chauvinistic functionalism, as it were. For example, Russell and Norvig (2003) complain that the Turing Test is unsuited as a benchmark for research in A.I., because it is not designed to test a system for its intelligence, but for its capacity to resemble human beings. As they put it, "aeronautical engineering texts do not define the goal of their field as 'making machines that fly so exactly like pigeons that they can fool other pigeons'" (ibid, p. 3). In other words, they reject TP for its bias towards a peculiar cognitive niche which lies at intersection between human behavior and intelligent behavior (Figure 2) – a niche which is fairly uninteresting from an engineering point of view, for at least two reasons. First, the Turing Test can be passed by systems which lack intelligence when judged by the normative standards of classical rationality (e.g., because of their susceptibility to logical fallacies, inaccuracy in arithmetical computations). Second, the Turing Test can be failed by systems because they succeed at cognitive tasks that outstrip the average intelligence of human beings (e.g., quickly solve difficult optimization problems).

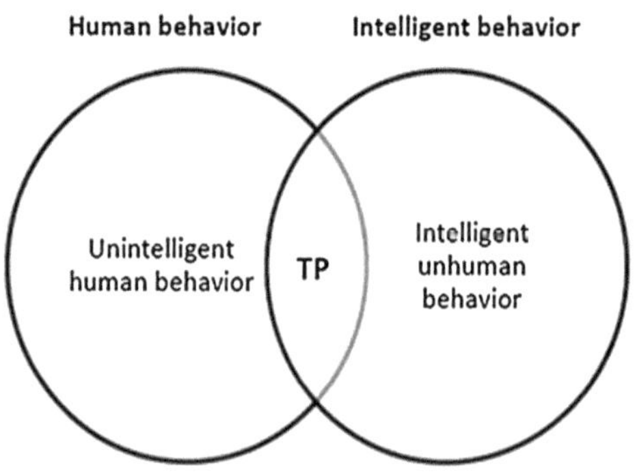

Figure 2. The cognitive niche carved out by Turing-Parity (TP)

It seems that in order to carve out a niche for extended cognition, an adequate understanding of Clark's PP requires a similar balancing act as a suitably constrained version of Turing's TP. First, both principles are incompatible with "neuro-chauvinistic" criteria of intelligence/cognition that are essentially tied to the exact neurophysiological substrate which underpins the cognitive functioning of our biological brains. Second, both principles are also incompatible with "chauvinistic" versions of functionalism, which require that cognitive states must exhibit the same fine-grained functional profile as their organic realizations inside the brain. Third, at least if we accept the criticisms of TP offered by Block and Dennett, we need to go beyond pure input-output equivalence by imposing further computational or temporal constraints (or both) on *how* the Turing Test ought to be passed. Similarly, Clark's commonsense functionalism ought to go beyond the "behaviorist" thesis that organism-bound and extended cognitive agents can in principle have the same behavioral dispositions.

There is ample textual evidence that Clark intends his interpretation of PP to go beyond mere input-output equivalence. For instance, Clark and Chalmers argue that the inscriptions in Otto's notebook must be understood as representational vehicles of his extended beliefs. This would seem to imply that the attribution of intentional mental states must be based on the operation of an appropriate representation-manipulating mechanism that is capable of mediating the effects of systemic inputs to behavioral outputs. Similarly, Clark (2001a, pp. 19f) has argued against the excessive liberalism of TP that we should not consider PARRY (Colby, 1975) – a "good-old-fashioned" symbolic A.I. program that relies on a few cheap hacks to mimic the verbal responses of a schizophrenic – as an intelligent cognitive system in its own right, even if it were somehow able to pass the Turing Test. This shows that Clark's functionalist understanding of PP is at least as strong as Blockian psychologism, and thus goes beyond TP.

Finally, as Russell and Norvig have argued, TP is skewed towards displays of intelligent behavior that we typically observe among ordinary human beings. In PP, we have the counterfactual comparison to unproblematic instances of inner human cognition which remains the gold standard that sets the bar for what else we ought to consider as cognitive. This excludes from the realm of cognition the ubiquitous and not distinctively *cognitive* forms of information-processing by electronic appliances like DVD-players or internet routers. Concluding our comparison between TP and PP, we can thus represent the peculiar niche of extended cognition that is carved out by PP as the region in which the realm of cognition and the realm of extra-neural information-processing overlap (Figure 3).

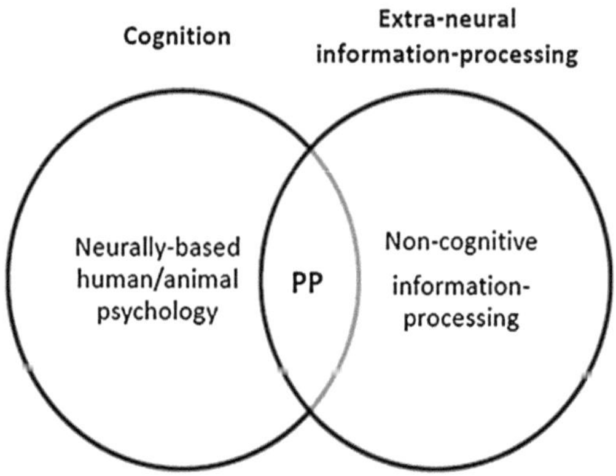

Figure 3. The niche for extended cognition carved out by Clark and Chalmers' Parity (PP)

We are now in a position to respond to Rupert's worry that PP threatens to turn the notion of cognition into a response-dependent concept. First, we have already granted that (esp. commonsensical) functionalism has an implicit "functional grain" parameter whose setting in particular contexts of explanation may affect the plausibility of extended cognition. However, beyond the familiar semantic problems caused by vagueness, it is not clear that functionalism turns the category of mental states into an ontologically response-dependent category.[19] Moreover, since this would be a problem for functionalism in general, we should plead EXTENDED FAIRNESS (Chapter 1.6) on this issue. Second, the criticisms of TP that we have reviewed suggest that the HEC-theorist should not be afraid to add further (e.g., computational, temporal, pragmatic…) constraints on *how* mental states ought to guide an agent's behavior "in ways that would not cause constant misunderstandings and upsets" (Clark, 2005a, loc. cit.) in ordinary communicative situations. There is no reason for a HEC-theorist to be a closet behaviorist and endorse an interpretation of PP that is based purely on gross behavioral dispositions.

2.7. In defense of double standards

The suggested relationship between TP and PP raises another puzzle that we now ought to address. To see this, consider Hewlett-Packard's famous first pocket calculator, the old HP-35. It came on the market at a time when most calculators had only the four basic arithmetical functions. The HP-35 greatly surpassed those models in terms of its functionality, speed, precision, and raw

19 Searle's (1980, 1992) critique of functionalism could be read as an argument along those lines.

computational power. Does this make the HP-35 an intelligent (albeit non-human) cognitive system in its own right? If we go by TP, the answer is clearly negative, because the HP-35 supports a far too limited range of intelligent conversations (or behaviors) to pass Turing's test (cf. Clark, 2001a, pp. 20-21). It would fail miserably at trying to respond intelligently to any questions outside the narrow domain of numerical calculations. However, as a portable, reliably available, trustworthy, and regularly employed resource in the hands of Otto-the-engineer – i.e., as a piece of "transparent equipment" (Heidegger, 1927/1977) – PP would seem to imply that Otto's calculations on his HP-35 should count as instances of extended arithmetical cognition. Is it coherent to defend such a "double-standard" about the cognitive status of one and the same physical system (the HP-35)?

In response, let us consider, first, an uncontroversial case of a non-cognitive extension. Let's say I decide to turn the Victorian terrace of my house into a multi-functional winter garden. By doing so, I create a spatial extension of my total surface of indoor living space. Conjoined with the rest of my house, the new winter garden adds significant value to my dwelling place that I can enjoy all year long. However, this does not mean that I would want to live within the confines of my winter garden. When considered in isolation, the lack of access to basic resources such as water, a bathroom, and a stove for cooking makes my winter garden as such an area that is hardly suited for habitation. More generally, the criteria that we apply to determine whether some room constitutes a minimal living space are different from those by which we judge whether an extension of the same basic room makes for an increase in effective living space.

A similar asymmetry, I want to argue, holds between the relatively persisting control architecture of the intelligent organism, and the more temporary, task-dedicated, yet potent forms of cognitive extensions that can be built on top of the former (cf. Clark, 2007; Clark, 2008, Chapters 5-6). As a way of fleshing out this functional asymmetry, Clark (2007, 2008) has recently proposed that we can paraphrase the HEC *salva veritate* as the

Hypothesis of Organism-Centered Cognition (HOC): "Human cognitive processing (sometimes) literally extends into the environment surrounding the organism. But the organism (and within the organism, the brain/CNS) remains the core and currently the most active element. Cognition is organism-centered even when it is not organism-bound" (Clark, 2008, p. 139).

From a HOC-perspective, the mobile human organism is best described as "the handy container of persisting recruitment processes and of a batch of core data, information, and body-involving skills. Thus equipped, [it] is revealed as a kind of walking BIOS, ever ready to bootstrap into existence the larger soft-assembled cognitive systems that are, quite literally, the information-processing engines of much advanced thought and reason" (Clark, 2008, p. 138). Given the operational asymmetry between the intelligent organism and its cognitive extensions, it is perfectly reasonable to impose different functional requirements which each of them ought to satisfy in order to count as (part of) a cognitive system. Those may include different levels of functional grain, which we have said affects the plausibility of extended functionalism.

Consider the functional requirements for a "mobile BIOS." It is not at all implausible to think that only certain specific types of information-processing strategies are intrinsically capable of supporting "fast thinking" (Dennett, 1987) – i.e., the kind of fluent, adaptive, efficient, pattern-sensitive sensorimotor engagement with the world that is the mark of intelligent behavior in the real world. Clark (2008, p. 107) has suggested that some sort of connectionist-style learning device may be practically indispensable to achieve the basic skills of perceiving, learning, and locomotion that human beings share with many other mammals. This fits an intermediate position that Clark (1989, pp. 35-36) once characterized as a kind of *micro-functionalism*. Micro-functionalism preserves a commitment to the multiple realizability of cognition by different physical substrates, but in addition, imposes the more fine-grained functional constraint that any nomologically possible substrate ought to support certain cognitively critical properties such as pattern recognition, flexible generalization, or graceful degradation (ibid.). According to Clark's micro-functionalism, this constraint severely constrains the space of computational architectures which are capable of producing and maintaining real-world intelligence (e.g., parallel distributed processing in neural networks).

But once the right kind of mobile, adaptive, plastic bundle of basic cognitive capacities is up and running, it can be enhanced with new kinds of computational and representational resources that may have little in common with the preexisting connectionist architecture. As "cognitive integrationists" have stressed (Chapter 2.5), the very fact that the manipulation of those bio-external resources supports hitherto unavailable forms of information-processing makes their integration into hybrid cognitive systems worthwhile. Consequently, extended functionalists should happily embrace a "double standard" theory of cognition. There is no reason to think that external resources must realize the exact same microfunctional properties that may be required for our biologically entrenched inner cognitive architecture in order to be combined as parts of an extended cognitive system.

2.8. Parity and coupling conditions

If the causal distribution of cognitively relevant labor deserves a proportional spread of cognitive credit – as accorded by PP – the boundaries of the mind become porous. For instance, do I believe the content of all *Wikipedia* entries that I can access online? Do I know the address of every butcher in town courtesy of *Google Maps*? If there is no way to distinguish between the genuine *incorporation* of external resources into one's cognitive system from more mundane cases of epistemic *tool* use, the HEC might seem to lead to a kind of pan-psychism ("cognitive bloat"[20]) that would trivialize its scientific import. In order to keep the uncontrolled proliferation of mental states under control, Clark & Chalmers (1998, p. 17; see also Clark, 2010, p. 46) posited a number of "coupling conditions" which a bio-external resource ought to satisfy in order to be admitted as part of an agent's extended mind.[21] The requirements were that

1. the external resource be reliably available and typically invoked when confronted with the task at hand,
2. any information thus retrieved be more-or-less automatically endorsed. It should be deemed about as trustworthy as something retrieved clearly from biological memory, and thus not usually be subject to critical scrutiny,
3. any information contained in the resource should be directly and easily accessible as and when required,
4. the information in the resource has been consciously endorsed at some point in the past, and indeed is there as a consequence of this endorsement.

As Clark and Chalmers argue, carefully applying these criteria can go a long way in curing excessive forms of cognitive bloat.[22] For instance, isn't the availability of the fast-rotate button in the original Tetris-example limited by the availability of the video console? Perhaps, but this wouldn't be a problem for a cybernetically enhanced player who uses neural implants to remote-control the on-screen rotations. Isn't the notebook too fragile and temporary to be part of Otto's extended mind? No, because knowing how vital the notebook is for his cognitive success, Otto makes every effort to shelter it from loss and harm –

---

20 This term is due to Clark (2001a, p. 156).

21 Although Clark and Chalmers did not use the term "coupling conditions" in their original paper, it has become standard usage in the current debate. For a brief discussion of the technical usage of "coupling" in dynamical systems theory, see Chapter 3.

22 The following examples are directly taken from Clark (2010, pp. 46f).

much like he cares about a vulnerable biological limb of his. Surely Otto usually puts down the notebook e.g., when he showers – but why should a temporary disconnection lessen the centrality it plays in Otto's mental life? What matters is that he is reliably disposed to use it whenever the cognitive demand arises. For instance, Otto won't admit ignorance on a subject before he has carefully consulted his notebook. Similarly, the fact that Inga does not actively entertain most of her beliefs while she is asleep (or at any given time during her waking hours, for that matter) does not undermine our practice of attributing those beliefs to her as dispositional mental states. Finally, Otto's notebook may be subjected to fraudulent intervention by malicious agents, but sinister neuro-surgeons might equally tamper with Inga's brain. In either case, the sheer possibility of epistemic sabotage does not defeat the actual reliability of Inga's memory when it operates veridically under normal circumstances, in ecologically normal environments.[23]

It is instructive to apply the suggested coupling conditions to the cognitive machinery that we take to be located *within* the brain, and thus part of a single cognitive agent. Against the plausibility of Clark and Chalmers' account, Rupert (2004, p. 424, fn. 58) has argued that many states of the brain that we would normally count as an agent's non-occurrent mental states would actually fail to satisfy at least one of the proposed coupling conditions. For example, some beliefs and memories do not readily come to one's mind even though their content would be relevant in the situation at hand, and despite our best efforts to retrieve them. On the other hand, sometimes people actively repress the memory of stressful or traumatic autobiographical events, or distrust their inner memories, even if they are in fact reliable, because they are uncertain about the source or the content (or both) of their memories. Does this refute, or at least render obsolete Clark and Chalmers' account of coupling conditions?

The first thing to point out in response is that the boundary of the skull is indeed not always a metaphysically unassailable guide to cognitive ownership, as Nagel's (1971) discussion of personal identity in split-brain patients has shown (see also Chapter 6.2). But more importantly, we should bear in mind that Clark and Chalmers' account of coupling conditions was never meant to provide a general criterion for demarcating *cognitive* from non-cognitive states across the board. Since it was designed to scratch a particular itch, i.e., cure the problem of cognitive bloat, its primary purpose is to demarcate *extended* cognitive states from extended, but non-cognitive resources. It does so by specifying a number

23 On the possibility of epistemic sabotage, and how it affects the theoretical distinction between perception and introspection, see Sterelny (2004b), and the reply by Clark (2008, Chapter 5.8).

of conditions that are jointly *sufficient* for a bio-*external* resource to be considered as part of one's mind. This does not imply that each condition is individually *necessary* for a bio-*internal* resource to be considered part of one's mind. In the previous section (Chapter 2.7), we have argued for a "double-standard" theory of the minimal functional requirements of biologically primitive internal vs. extended cognitive states. In similar vein, it is open to the HEC-theorist to argue that coupling conditions should be weighed somewhat differently for bio-internal and bio-external resources, or that they ought to be resolved with respect to different standards depending on the context of folk-psychological explanation.

Moreover, the possibility of counterexamples that are contrived to upset our vernacular intuitions about the mind does not automatically undermine the explanatory value of the HEC. This is because first, counterexamples are not necessarily lethal for a theory unless they crop up too frequently, and in areas that are central to the intended domain of its application. But for many central cases, the judgments formed on the basis of the proposed criteria do not stray too far from our commonsensical intuitions. Second, as we have said earlier, HEC-theorists should stick to their guns and not be afraid of a little revisionism concerning our received, commonsensical understanding of mental states. This is particularly true of folk-psychological attributions of mentality that are tainted with the "Cartesian" presupposition that the domain of cognition is more or less identical with the domain of consciously occurring mental states.

Although I have vowed to bracket the problem of consciousness, I need to make an exception here in order to address the controversies over the fourth coupling condition, the *past-conscious-endorsement* criterion. The past-endorsement criterion is the only coupling condition which explicitly refers to a conscious mental act. One might speculate that Clark and Chalmers originally introduced the fourth condition to tighten up their account of functional coupling, as a way of patching all the gaps which the other three conditions had left open. However, it should be pointed out that Clark and Chalmers did not fully endorse the fourth condition in the original paper, and it has silently disappeared from Clark's more recent presentations.[24] This might simply be due to a lack of space or other, purely pragmatic reasons. However, there are reasons to think that its disapperance is calculated, because it does not sit particularly well with the spirit of HEC. An example discussed by Rupert (2004, pp. 402-405) will help to flesh out this suspicion.

A very large number of people today own a cellular phone, which gives them almost instantaneous access to an automatic telephone directory service. If

---

24 E.g., only the first three criteria are listed in Clark (2010, p. 46).

we go by the first three coupling conditions, it would seem that every adult who regularly consults the telephone directory via her cell phone has (mostly) true beliefs about everybody else's phone number listed in that directory. This is because the directory assistance operator is a relatively permanent and easily accessible source of information, and Otto typically trusts what the operator tells him more or less automatically. If the suggested conclusion sounds counter-intuitive to our ears, what resources does the HEC-theorist have to winnow extended mental states from all the other innumerable sources of information to which an average user of smart phones can at least potentially access in our hyper-connected information environment? Surely something like the past-endorsement criterion would come in handy at this point.

But as Rupert (ibid.) has argued quite convincingly, it would be ill-advised for the HEC-theorist to let the past-endorsement criterion do the required heavy lifting. For one, it would suggest that history is at least partly constitutive of belief-ascriptions (see Clark & Chalmers, 1998, p. 17, fn. 5). This does not sit well with the assertion that according to the HEC, "the relevant external features are *active*, playing a crucial role in the here-and-now. Because they are coupled with the human organism, they have a direct impact on the organism and on its behavior" (Clark & Chalmers, 1998, p. 9). The active-role clause distinguishes HEC from traditional versions of content externalism (Chapter 1.1), where distal environmental features that account for mental differences between otherwise identical hypothetical "twins" tend to be "dangling at the other end of a long causal chain" (ibid.). Second, as Clark and Chalmers (ibid.) point out, subjects can acquire ordinary internal beliefs through processes of which they do not have any conscious awareness, e.g., due to subliminal perception or memory tampering. Moreover, it would seem to be at least metaphysically possible that people acquire new beliefs as the result of telepathy or clairvoyance. Applying the past-endorsement criterion would rule out these cases of non-extended beliefs on apriori grounds, which seems unsatisfactory. Third, it would arguably go against the grain of the externalist project of jettisoning the Cartesian vision of an "inner homunculus" as the true subject of mental states (Rupert, 2004, p. 404). If we demand that an external resource – even one that is poised to play the right kind of functional role – needs to be consciously endorsed to get the seal of mental authenticity, this would seem to be a fairly strong concession to "Cartesian" internalist intuitions.

Clark's decision to scrap the past-endorsement criterion may thus be well-advised. But what can we put in its place to block the cell phone user's extended cognizance of everybody else's phone numbers? Of course we could simply bite the bullet, and chalk up the so-called counterexample to antiquated intuitions that need to be overhauled in this day and age of almost ubiquitous, high-speed mobile internet access. Alternatively, one could argue that the act of conscious endorsement is itself an extended cognitive process. However, unless we have independent criteria for demarcating the bounds of extended consciousness, this would not really solve our problem. If the act of conscious endorsement is extended, on what grounds does Otto lack an extended conscious cognizance of everybody else's phone numbers?

A more promising move for the HEC-theorist, suggested by Rupert himself (2004, p. 403, fn. 26), is to require that the biologically-bound cognizer must at least be *causally responsible* for producing whatever external pieces of information can rightfully be considered part of her extended memory.[25] Such a causal-responsibility criterion might be permissive enough to let in Otto's notebook, but rule out an anonymously created and publicly available telephone directory. But Rupert (ibid.) immediately retorts that such a criterion would still be too liberal unless we also demand that the relevant information must somehow be consciously grasped. This is because a smart phone user can generate a huge database of phone numbers with a few key strokes by submitting a query in online telephone directories. In response, Rupert (ibid.) suggests that the HEC-theorist might further require that the retrieved phone numbers must be *personally tailored* to the information-seeking agent in order to be considered as part of her extended mind.[26] Suppose Otto submits a query which returns an enormous amount of telephone numbers. Even though Otto is causally responsible for typing and submitting the query, only a tiny fraction of the retrieved results is what Otto was actually looking for. Intuitively, the irrelevant search results are not disposed to guide Otto's behavior, thus they should not count as part of Otto's extended beliefs.

I think that a suitably qualified combination of the causal-responsibility and the personal-tailoring requirements could be used to tighten up Clark and Chalmers' account of coupling conditions, although it would most certainly not avoid all possible counterexamples. Let me spell out a few suggestive remarks, with the help of some ideas from the science of *information retrieval* (Baeza-Yates & Ribeiro, 1999). For example, consider a user who searches for information in a database through a customizable search engine. The search engine ranks the items of the database in terms of an underlying model of how likely they match the user's query, and returns only the top-ranked items. This outcome is achieved by assigning a degree of relevance to every item in the database, which is based on its estimated likelihood of satisfying the user's informational needs. Importantly, for an item to be considered as *relevant* for a certain type of user, it

---

25 Rupert (loc. cit.) attributes this response to William Lycan (pers. com.).

26 See also Clark (1997, p. 217) for a proposal along these lines.

is not necessary that the particular user submitting the query must have previously endorsed the item in question. The degree of relevance is determined instead by a generic model of anticipated user preferences that is built into the information retrieval system. This would still introduce some historical dependency into our story, because user models are partly constructed on the basis of a user's past behavior, together with information about general tendencies of the past behavior of similar users. However, this historical dimension is not incompatible with a purely dispositional reading of the personal-tailoring criterion. It concerns only the mechanics of the algorithm which the search engine uses to infer the customized user models, not the question of whether the result of a particular query is objectively disposed to satisfy a user's thirst for information *here-and-now*.

A variety of different measures for evaluating the performance of information retrieval systems have been proposed in the literature (e.g., Baeza-Yates & Ribeiro, 1999). Among the most basic measures are the *precision*, *recall*, and *fall-out* of a query (with respect to a collection of data). *Precision* is the probability that an item retrieved by a query is relevant for a given user. *Recall* expresses the probability that an item which is relevant for a given user is in fact retrieved by a query. The *fall-out* refers to the proportion of false positives, i.e., the probability that a non-relevant item is retrieved. These three measures are interrelated. For instance, it is trivial to achieve perfect recall simply by indiscriminately returning every stored item to a query, but this would clearly hurt precision and produce a large fall-out. Conversely, by returning zero items in response to a query, it is possible to avoid any fall-out, but obviously at the expense of recall. If Otto seeks out information in a completely generic search engine, his query is bound to return items with low precision, low recall, and high fall-out. Since the retrieved information is insufficiently tailored to Otto's specific needs, it should not be considered as part of his extended beliefs, memories, (or desires, in the case of suggestions by a recommendation engine). On the other hand, if Otto's query returns items with very high precision, very high recall, and very low fall-out, the personal-tailoring criterion would imply that the information they contain can plausibly be considered as part of Otto's extended cognitive system. To deal with intermediate cases, one could contrive more sophisticated versions of the personal-tailoring criterion, perhaps expressed as weighted functions of various performance measures.

In the end, it is most plausible to think that the membership of bio-external resources in a cognitive system runs alongside a continuum (cf. Wilson & Clark, 2009). This means there are bound to be cases for which there may not be a determinate fact of the matter as to whether we should count them as part of somebody's mind. For those borderline cases, ascriptions of extended mental states would remain irreducibly vague (cf. Clark & Chalmers, 1998, p. 17). Several philosophical approaches have been devised to deal with the semantics of vague predicates. For instance, a supervaluationist approach (Fine, 1975) would grant that some ascriptions of extended cognition are neither true nor false, but alternative precisified versions (e.g., depending on our exact statement of the coupling conditions) would have a definite truth-value. On the semantics of vagueness, I plead FAIRNESS and EXTENDED FAIRNESS.

As an afterthought, let me point out that our discourse about cognitive ownership may be far more sensitive to eminently practical concerns (such as portability, or information retrieval) than philosophers have traditionally assumed. Just consider the peculiar turn which our discussion took in the present section. We began by asking the seemingly factual question: *what does it take for an environmental resource to be considered as part of somebody's mind?* But we ended up with an engineering problem on our hands: *how can we build information retrieval systems that better allow human beings to extend their minds?* This turn of events is perhaps indicative of what the next generation information technologies might have in store for us: the increasingly seemless integration of smart phones and portable digital assistants into our problem-solving repertoires, the impending arrival of "ubiquitous computing" (Weiser, 1991), and the near-future availability of neural implants are prone to smear the boundaries of cognitive ownership even further (Clark, 2003). Understanding this development will require unified theoretical perspectives in which the flow of information retrieval can be studied both within and between organic and artificial computational systems (Norman, 1999; Clark, 2003; see also Clark 2008, pp. 118-22).

2.9. Parity and scientific kinds

Conceiving of the HEC as a "natural progression in the maturing of the sciences of the mind" (Clark, 2008, p. 219), Clark has praised its virtue as a fertile and illuminating framework which allows us

> [...] to see mind and intelligence themselves as mechanically realized by complex, shifting mixtures of energetic and dynamic coupling, internal and external forms of representation and computation, epistemically potent forms of bodily action, and the canny exploitation of a variety of extrabodily props, aids, and scaffolding" (p. 219).

To ensure steady progress along the projected path, Clark recommends a rich blend of methodological pluralism:

> "To unravel the workings of these embodied, embedded, and sometimes extended minds requires an usual mix of neuroscience, computational, dynamical, and information-theoretic understandings, 'brute' physiology, ecological sensitivity, and attention to the stacked designer cocoons in which we grow, work, think, and act" (p. 219).

Is this kaleidoscopic vision of an embodied, embedded, and sometimes extended cognitive science a recipe for success or disaster? In particular, does the emergence of our "mashup" minds "from a dizzying motley of heterogeneous elements and processes" (ibid.) offer clear theoretical objectives, and a sound methodology for the scientific study of cognitive phenomena? It is often considered a "mark of the scientific" that science strives to give causal explanations that carve nature at her joints. Using the words of Adams and Aizawa (2001), this means that

> "science tries to get beneath observable phenomena to find the real causal processes underlying them; science tries to partition the phenomenal world into causally homogeneous states and processes. Thus, as sciences develop a greater understanding of reality, they develop better partitions of the phenomenological" (ibid., p. 51).

According to Adams and Aizawa, the status of psychology as an empirical science requires that cognitive states and processes are taxonomized in terms of underlying causal mechanisms. This project would be seriously undermined by the HEC-theorist's proposal to treat a gerrymandered bunch of neural, bodily, and bio-external resources as parts of cognition. This is because bodily and environmentally extended processes would be sustained by causally heterogeneous mechanisms between which the "differences are so great that they can be seen not to form a cognitive kind" (ibid., p. 52). To remain (or perhaps become) a real science, psychology should therefore leave the study of trans-cranial information-processing to other, non-psychological branches of science and engineering.

In response to this worry, let us consider what is meant by the causal heterogeneity of extended cognitive processes, and why it supposedly undermines the inductive fertility of cognitive kinds. For starters, the problem may simply be that

> "the causal processes in the brain appear to be nothing like the firing of electrons at a phosphorescing screen [which occur when we rotate Tetris blocks on-screen], generating marks on a paper through friction with a graphite shaft, generating marks on a paper with a ball point pen [...]" (ibid., p. 57).

But citing "brute" causal differences alone hardly carries any theoretical weight. It is beyond controversy that the physical mechanisms underlying internal and extended cognition involve different media, but this does not rule out the possibility that there are interesting information-processing commonalities between physically diverse mechanisms. For instance, generating marks on paper through friction with a graphite shaft, generating marks on paper with a ball point pen, and typing letters on a keyboard are all instances of *writing*. My PC and your Mac are two physically distinct machines, but there is a computational level of description at which they can be said to share a von Neumann architecture, or implement a Universal Turing Machine (Herken, 1995). Finally, even the orthodox computational theory of mind asserts that biological brains and computers can implement the same cognitive architecture – such as the classical physical symbol systems-based cognitive architecture to which Adams and Aizawa are committed (ibid., p. 49; see also p. 60). If Adams and Aizawa want to stay within the minimal functionalist consensus, they ought to at least grant that cognitive states and processes are multiply realizable in principle. EXTENDED FAIRNESS (Chapter 1) further dictates that a plausible understanding of natural kinds that can reasonably be presupposed in this debate should be able to accommodate the explanatory needs of classical cognitive science. This excludes an appeal to metaphysically sparse notions of natural kinds that classify things purely in terms of their physico-chemical structure (Kripke, 1980; Putnam, 1975b), or "perfectly natural" microphysical properties (Lewis, 1983). According to a suitably liberalized understanding, any predicates that support successful inductions can potentially express natural kinds, and this includes those to which the special sciences refer in their law-like generalizations (Fodor, 1974; 1997; Kornblith, 1993; Boyd, 1999).

Being sympathetic to functionalism, Adams and Aizawa (2008, p. 69) have recently clarified that they are ready to concede that in principle, mental states could be realized by non-organic resources. More specifically, they admit that transcranial cognitive processes that are spread out across neural and extra-neural resources are metaphysically and even nomologically possible. What they deny is that transcranial cognitive processes are actually observed in mundane instances of human cognition. In what follows, let us therefore focus on their specific complaint that the putative range of extended cognitive processes would be too gerrymandered to support any interesting law-like psychological generalizations. For instance, whereas biological memory systems "display a number of what appear to be law-like regularities, including primacy effects, recency effects, chunking effects and others" (Adams & Aizawa, 2001, p. 61), there is no reason to believe that those effects generalize across all types of external mnemonic devices (such as bulletin boards, photo albums, or iPhones). Consequently, they conclude that

> "transcranial processes are not likely to give rise to interesting scientific regularities. There are no laws covering humans and their tool use over and above the laws of intracranial human cognition and the laws of the physical tools" (Adams & Aizawa, 2001, p. 61).

In support of the same conclusion, Rupert (2004, pp. 406-415) reviews a large body of literature on short term, long term, and working memory, cataloguing their characteristic properties such as learning-time, access-time, and the occurrence of interference effects during memory recall. Noting that practically none of those features are displayed by external memory resources, he asks:

> "[...] why think that the apposite external and internal states (or forms of processing) are thereby of the same causal-explanatory kind? Why infer the existence of one overarching kind, memory, subsuming both internal and external states and processes, that will be of significant explanatory use in cognitive science?" (Rupert, 2004, p. 409).

Of course it would be foolhardy to anticipate the results of future research, predicated on the HEC, to speculate whether generic cognitive kinds which subsume extended and non-extended forms of cognitive processing will eventually earn their place in the maturing sciences of the mind. The proof of the pudding is in the eating, as they say. But from a more philosophical perspective, we can at least examine the general grounds on which the critics' inductive skepticism about extended cognitive kinds seems to be based. If there is evidence for the inductive success of physically heterogeneous, generic kinds in other domains of science, this would at least help attenuate the critics' skepticism about the causal-explanatory value of the HEC.

Adams & Aizawa's assumption that "the cognitive must be discriminated on the basis of underlying causal processes" (ibid., p. 52) rests on a specific conception of causal-explanatory kinds. According to their conception, respectable forms of explanatory unification require the *causal homogeneity* of the underlying mechanisms by which higher-level regularities are realized; second, they require that all elements of the subsumed domains behave according to the exact *same laws* (cf. Clark, 2010, pp. 50-52). Both assumptions, I shall argue, do not mesh well with our actual scientific practice.

First of all, causal heterogeneity already begins at home, in the brain. As Clark (ibid.) has pointed out, the anatomical boundary of the skull creates the false impression that the cognitive kinds that are recognized by our familiar folk-psychological taxonomies are causally homogeneous. But appearances can be deceiving. For example, consider the case of vision. There is evidence that the neural mechanisms responsible for conscious seeing (e.g., visual recognition, categorization) are different and operate independently of the neural mechanisms that support non-conscious forms of visuo-motor processing (such as visually guided reaching; cf. Milner & Goodale, 1995; Goodale & Milner, 2004). Similarly, the neural correlate of action-related mental imagery (e.g., watching sports, or imagining how to type a sentence) involves the activation of the associated motor responses, whereas others (e.g., imagining a lake) do not (Decety & Grèzes, 1999). Certain aspects of visual processing rely on an equivalent of look-up tables, while others do not (Churchland & Sejnowski, 1992). Nevertheless, there is an important sense in which all these mechanisms, despite the heterogeneity of their underlying neural signatures, contribute in one way or another to the visual functioning of an organism. In sum, it is by no means evident that the causal homogeneity *within* internal cognitive processes is significantly larger than between internal and extended cognitive processes.

Second, and moving from defense to offense, the inductive power of grouping together superficially dissimilar systems is well-documented by the vast literature on complex systems (Gell-Mann, 1994; Ball, 1998). While those systems are composed out of very different materials, they share interesting scale-invariant patterns of organization. For example, in many systems, individual elements move around randomly. If a moving element touches another element, it becomes attached. The emergent result is a fractally connected branching aggregate. This process, called *diffusion-limited aggregation* (DLA), has been implicated in the growth of human lungs, frost on glass, and cities (Ball, 1998). Creating a category of DLA phenomena is scientifically useful because the resultant overall patterns of thin, fractally connected branches, are remarkably similar, and have almost identical statistical properties. Once changes in complex arrangements of matter are classified as examples of DLA, one can better predict how they will change with time, how they will be affected by temperature, what shapes they can and cannot attain, and so on. Physicists often use the term *universality* to describe generic phenomena (such as DLA) which happen in the same way across a wide range of different kinds of physical systems. The discovery of similar universality classes does not require the functional homogeneity of the underlying causal mechanisms, nor does it entail that all elements of the class behave according to the exact same laws.

Third, scientists are currently exploring potential analogies between extended cognition and extended life that would point towards further explanatory convergence between the cognitive and the life sciences. As we have said earlier, central to the HEC is the idea that human beings systematically modify their environment to enhance their cognitive capacities – that "thinking is a kind of building, a kind of intellectual niche construction," as Wilson and Clark (2009, p. 58) aptly put it. At the same time, the theoretical motif that organisms gainfully manipulate, appropriate and integrate environmental resources into their pre-existing organic structures in order to change their biological conditions has gained a strong foothold in the life sciences. There are now several theoretical frameworks –notably *niche construction theory*, *developmental systems theory*, and the idea of *extended physiologies* — that place a renewed emphasis on the importance of environmental "scaffolding" (Wilson & Clark 2009; see also Wilson, 2004, 2005).

First, the activity of niche construction has been re-discovered as a "neglected process" (Odling-Smee, Laland, & Feldman, 2003) in evolution. Until fairly recently, evolutionary biology was dominated by the orthodox Darwinian analysis in view the environment is viewed as a fixed set of constraints which determine an organism's fitness. This assumption has been most visible in the construction of optimality models. For instance, consider the problem of how a bird feeding its young should trade the number of trips back to the nest against the load carried on each trip. What strategy will maximize the rate of food delivery to its offspring? Answering such questions, a theorist has to specify the set of strategies that are available to an agent, and determine the costs and benefits of each strategy against the backdrop of a given environment. The purpose of optimality models is then to figure out the best trade-off between those costs and benefits, and use it to analyze the evolutionary dynamics of the target environment.

Whatever "tactical" advantage can sometimes be gained from this idealizing assumption, it is clear that in the real world, organisms typically do not just passively respond to selection pressures dictated by their environment. In many cases, organisms themselves modify those selection pressures by filtering and transforming the physical challenges which the environment imposes on their bodies. According to Odling-Smee et al. (2003), the process of niche construction "occurs when an organism modifies the feature-factor relationship between itself and its environment by actively changing one or more factors in its environment" (ibid., p. 43). For instance, part of what makes beavers so well-adapted to forage for resources in their surrounding environment is their ability to build dams. First, because dams enlarge the shoreline and thus reduce its seasonal variability, they increase the supply of food trees. Second, in virtue of engineering a stream, the presence of dams also allows beavers to cover greater distances by water – a path that is safer than traveling by land, and on which it is easier to transport food. If the fitness-enhancing effects of these environmental alterations remain stable across many generations, the niche-constructing behavior itself becomes visible to natural selection and thus subject to evolution. A central theoretical goal of niche construction theory is to integrate ecology and genetics. Corresponding to the traditional organism-based notion of *genetic* inheritance, niche construction theorists argue that traits can also be inherited through the transmission of modified environments – the idea of *ecological* inheritance (Odling-Smee, 2010).

Second, in response to the dominant gene-centered view of ontogenetic development, proponents of developmental systems theory view entire developmental systems as the basic units of developmental analysis (Oyama, 1985; Oyama, Griffiths, & Gray, 2001; Jablonka & Lamb, 2005). While genes are certainly one important resource by which nature constrains the development of organisms, they are only one element in a complex matrix of causes and constraints that factor into a developmental system. Epigenetic systems, such as the chromatine marking system, extend beyond the nucleus of the cell; other mechanisms for shaping the development of an individual's behavior, such as culture, extend beyond the organism into its physical and social environment. By working with an extended notion of *development*, developmental systems theorists seek to overcome traditional dichotomies such as nature versus nurture, and biology versus culture.

Third, building partly on Richard Dawkins's (1982) notion of the *extended phenotype* (which includes e.g., the web beyond the spider's body), the ecological physiologist Scott Turner (2000) has studied a range of physiological processes – e.g., metabolism, energy transfer, homeostasis – which extend beyond the boundaries of the animal's outer membranes (e.g., its skin, shell, soft tissue). For instance, Turner has shown that important component processes of a polyp's physiology are distributed over the coral reefs which the polyp inhabits, such as their calcite deposits and protozoan flagellates. Termite mounds, to cite another one of Turner's favorite examples, are animal-built devices for capturing wind energy whose physiological function is to power the active ventilation of the nest atmosphere. According to Turner, they are best viewed as accessory organs of gas exchange which serve the respiratory needs of the subterranean termite colony located beneath the mound.

Underlying these frameworks is the idea to view biological inheritance, ontogenetic development, and also physiology as constructive processes by which organisms "scaffold," and thus enhance the functionality of their internal organic structures. They all provide evidence for what Wilson (2005, 2006) has dubbed the *hypothesis of extended biology* (HEB), based on an analogical extension of PP to the realm of biology:

BIO-PARITY: If, as we confront some task, a part of the world functions as a process which, *were it to go on within the outer membranes of the animal*, we would accept as part of the biological process, then that part of the world *is* (for that time) part of the biological process.

The central message of the HEB is that there is nothing sacrosanct about the hull of the organism for the scientific study of biological processes. What remains to be seen is how far we can take the suggested analogy between the HEB and the HEC. Wilson (2006) distinguishes between a weak and a strong inference based on this analogy. According to the weaker version, an extended view of biological processes *encourages* the adoption of a corresponding view of cognition. According to the stronger version, the full understanding of a whole range of biological processes *requires* an extended view; consequently, the same unit of analysis is mandatory for our understanding of cognitive processes. The stronger argument would allow us to directly infer the HEC from the HEB. For present purposes, the weaker version is enough to show that we ought to take seriously the idea of hybrid, and causally heterogenous mechanisms which stretch beyond the organism's skin into the world.

Finally, let us conclude our survey by revisiting Adams & Aizawa's (2008, Chapter 4) more recent justification of their "taxonomy-by-mechanism" approach to the individuation of cognitive processes. Compared to the totality of physical and chemical processes which pervade our universe, cognitive processes are a much rarer occurrence in space and time. In order to explain the relatively sparse distribution of cognition, Adams & Aizawa propose that

> "one would expect that processes as distinctive as cognitive processes would supervene on correspondingly distinctive lower level processes. Roughly speaking, lower-level processes should be as distinctive as the higher-level processes they realize" (ibid, p. 68).

Following Wheeler (2010a, p. 250), let us call this assumption the *Distinctiveness Principle* (DP). The point of DP is to justify why we ought to care about the fine-grained details of human neural information-processing for the business of demarcating the bounds of cognition. As an example of DP, Adams and Aizawa cite the categorical differences between neural and extra-neural mechanisms involved in low-level vision. Before incoming light hits the retina, the propagation of photons within the optical apparatus of the eye is governed by the same basis laws of reflection, refraction, etc. that we find in the entire physical world. But once visual stimuli are "transduced" into neural impulses by the distinctive molecular structure of the retinal machinery, fundamentally different kinds of perceptual processes begin to take place that appear to have no functional equivalent in the extra-neural world. According to DP, the retina thus constitutes a non-question-begging mechanistic interface to demarcate the distinctive range of perceptual and cognitive processes from the far more ubiquitous occurrence of optical processes (Adams & Aizawa, 2008, p. 69).

However, DP does not seem to hold across the board for higher-level causal-explanatory kinds. A striking example discussed by Wheeler (ibid, citing Doolittle, 1994) is the occurrence of *functional convergence* in evolution. Functional convergence refers to an evolutionary process by which two (or more) biological entities come to perform the same function, but do so by way of entirely different underlying structures and mechanisms (e.g., because the relevant biological traits have evolved independently in each lineage). Wheeler cites the convergent evolution of alcohol dehydrogenases, i.e. a group of enzymes that occur in many organisms, and whose function is to break down alcohols that would otherwise be toxic. Even though alcohol dehydrogenases perform functionally equivalent roles for (say) vertebrates and fruit-flies, the respective enzymes have dissimilar chemical structures, and catalyze different chemical reactions to achieve the same outcome. If it can be shown that DP does not generally hold for biological kinds, there is no apriori reason to think that psychological kinds would satisfy DP. Consequently, DP cannot be used to undermine the plausibility of extended functionalism and the HEC.

2.10. Parity and intentionality

Adams & Aizawa (2001, 2008, 2010) have raised another important theoretical challenge to the HEC that is not directly tied to the issue of functional similarity between inner and outer resources. As critics of the HEC, they accept the *epistemological* role of PP (cf. Chapter 2.2) insofar as they accept that the burden of proof is on the internalist to show why the biological boundary of the organism – in particular, of the brain – should matter for demarcating the realm of the cognitive. According to them, the distinctive "mark of the cognitive" is its *intentionality* ("aboutness," "meaning"):

> "More specifically, it [= PP] means that being inside the brain cannot be the mark of the cognitive. This seems to us true and obvious. The bounds of cognition must be determined by finding the mark of the cognitive, then seeing what sorts of processes in the world have the mark. Following this method, we see that, as a matter of contingent fact, the cognitive processes we find in the real world all happen to be brain-bound. It appears to be just a contingent empirical fact that cognitive processes are not transcorporeal processes" (Adams & Aizawa, 2001, p. 46).

What distinguishes genuine cognitive from non-cognitive processes – such as digestion, the patellar reflex, and non-cognitive forms of information processing by video-game consoles – is that cognitive processes operate on mental representations which are intrinsically meaningful. In contrast, external symbolic artifacts such as traffic signs, newspaper headlines, and the scribblings in Otto's notebook have no intrinsic meaning. They "mean" what they do only because people assign them an interpretation. Whatever "meaning" external symbolic structures have is conventionally derived from the interpretative practices of the people who use them. Environmental resources thus do not realize genuine cognitive states, because they lack intrinsic intentionality. Based on their suggested mark of the cognitive, we ought to reject the HEC, because "whatever is responsible for non-derived content seems to find a place only in brains" (ibid, p. 63).

How exactly should we understand their talk of a "mark of the mental"? Describing their position, Adams and Aizawa (2001) say that they are committed to the view that non-derived intentional content is at least a *necessary* condition for a state or process being cognitive (p. 48). Hence we can cast their attack against HEC as the following argument:

1. Any resource that is mental has intrinsic content.
2. Any resource that has intrinsic content is found only in brains.

Therefore, any resource that is mental is found only in brains.

The distinction between intrinsic and derived meaning has a venerable tradition in philosophy which stretches back into ancient philosophy of language. In the *Cratylos*, Plato (1997) famously pondered the question of whether the relation between language and things is grounded in nature (*physei*) or is a matter of conventional stipulation (*thesei*). For instance, whereas it is a matter of natural law that smoke indicates the presence of fire, it appears to be a matter of convention whether one uses the English word 'fire' or the German word 'Feuer' to denote the presence of fire. According to common lore, Aristotle's answer to Plato's problem has been to invoke *concepts* as intermediary mental entities grounding the otherwise arbitrary relation between words and things. There is a considerable intuitive appeal to the Aristotelean view. If we think about fire, it does not seem to be a matter of convention that we're thinking about fire rather than (say) water. The conventional aspect of language only seems to affect our choice to use different words denoting one and the same concept. The relationship between thoughts and their objects is what the neo-Aristotelean Franz Brentano (1874) called *intentionality*. In contemporary philosophy of mind, "Brentano's problem" is to explain what it is for thoughts to be *about* something else – i.e., to stand for, or to represent certain things, properties, or states of affairs.

Until roughly the 1960s, there was a broad consensus among mainstream analytic philosophers that any psychological explanation which appeals to the intentionality of mental states is incompatible with physicalism. The reason for this perceived incompatibility was the apparent failure to translate our intentional vocabulary into a purely physical, non-intentional vocabulary, thus barring a successful reduction of the former to the latter (Chisholm, 1957; Quine, 1960). Consequently, Quine (1960) raised the dilemma that we either have to reject a physicalistic ontology in order to preserve the veracity of our intentional idiom; or we accept physicalism, but are thus forced to eliminate the intentional idiom as devoid of content.

But the philosophical tide has shifted over the last forty-something years. The reigning materialist orthodoxy now rejects Quine's dilemma, maintaining that it is possible to provide a "naturalistic" solution to Brentano's problem. A theory of intentionality is said to be naturalistic if it is consistent with the thesis that the fundamental inventory of the universe is nothing but what is described by the basic physical sciences. Since intentionality would thus not count as a fundamental feature of the universe, the main goal of the naturalization project is to show how exactly intentionality can be understood in materialistically kosher terms. In short, the name of the game is to show how to "bake a mental [intentional] cake using only physical yeast and flour" (Dretske, 1981, p. xi). An influential attempt at naturalizing the intentionality of mental states is known as the representational theory of mind (Fodor, 1975, 1987; Dretske, 1981, 1991; Millikan, 1984; 1989). According to this theory, a subject's thought that dogs bark involves her having a mental representation *R* of barking dogs, and the substance of the competing proposals consists in various accounts of the non-intentional facts which determine that *R* represents dogs barking (rather than, say, cats meowing).

Adams and Aizawa's criterion of intrinsic intentionality as a mark of the cognitive is premised on their acceptance of a "naturalistic" representational theory of intentionality (2001, pp. 48-51). Hence their notion of intrinsic representational content differs in important respects from John Searle's (1980, 1984, 1992) notion of intrinsic intentionality. Searle also conceives of intentionality as a biologically primitive feature of mental states that is produced by the brain, but further holds that intentional mental states must be at least potentially accessible to consciousness. Hence it would appear that for Searle, availability to consciousness is really the ultimate mark of the mental. Adams and Aizawa, on the other hand, readily grant that "qualia and phenomenal consciousness, while real and associated with cognitive processes, are not in themselves essential elements of cognitive processes" (2001, p. 48). In what follows, I shall therefore put aside stronger views which hold that consciousness is an essential ingredient of all intentional states (Searle, 1992; Strawson, 1994; Horgan & Tienson, 2002).

Contrary to Searle, Adams and Aizawa also do not claim that bio-external resources are nomologically incapable of producing intrinsic intentionality. Their opposition to HEC is not due to a brute "neuro-chauvinistic" (Block, 1978) bias. They readily concede that extended cognition is "logically and perhaps nomologically possible" (ibid., p. 47), and it is only "as a matter of contingent fact, [that] the cognitive processes we find in the real world all happen to be brain-bound" (Adams & Aizawa, 2001, p. 46). Accordingly, Clark's (2005a, p. 4) functionalist counter-proposal to take whichever neural ensemble passes Adams and Aizawa's test for intrinsic content, and then replace the entire structure, part by part, with functionally equivalent silicon devices, is not necessarily incompatible with their position.

In his reply to Adams and Aizawa's challenge, Clark (2005a, p. 4) initially raises the worry that the notion of intrinsic content may turn out to be "essentially fuzzy and indistinct" (ibid.), thus making it hard to find a firm enough common ground on which arguments from intrinsic intentionality can be evaluated. But the basic idea behind the distinction between intrinsic and derived content, I would argue, is fairly clear. We have a strong pre-theoretical grasp of what it is to be in a mental state with a definite mental content (e.g., the belief that I'm hungry right now). On the other hand, we stare in puzzlement at a menu written in a foreign language because we do not understand what it means (Searle, 1980). After having granted the intrinsic/derived distinction, Clark goes on to paraphrase Adams and Aizawa's objection as the worry that external representations are too "passive" and "parasitic, in their meaning, on public practices of coordinated use" (2005a, p. 5) to count as intentional mental states in their own right. A few lines later, he talks of external symbols as being "dormant" in Otto's notebook, waiting to be "awakened" by the cogitations inside Otto's head.

To counter Adams and Aizawa's objection *as understood by Clark*, Clark retorts that any plausible criterion of intrinsic content does not exclude the inscriptions in Otto's notebook once we distinguish between *occurrent* and *dispositional* mental states. Defending his version of extended functionalism, Clark thus argues that the entries in Otto's notebook are poised to guide Otto's actions in roughly the same way as ordinary dispositional beliefs (Chapter 2.3). The supposedly "dormant" meaning of the inscriptions in Otto's notebook merely reflects the causal-temporal dynamics of extended memory storage and retrieval. It does not show their lack of genuine representational content, provided that they are "appropriately linked, at run-time, to representations whose content is [...] intrinsic" (Clark, 2005a, p. 5).

Clark (ibid.) illustrates this point with several examples. He considers a Martian who is capable of storing large chunks of text in his brain in the form of bit-mapped visual images. Since the data are stored "passively" in the form of visual images, they cannot serve the Martian for tasks which require parsing and analyzing the stored data *as* linguistic representations. But we are further asked to imagine that the Martian can perform something like optical character recognition, and subsequently recover the linguistic structure of the text as a piece of meaningful prose. If intentionality is essentially a "potency" predicate in Clark's sense, it would be arbitrary to say that before the conversion from image to text, the bit-mapped representation of the image would not count (at least in a dispositional sense) as a *mental* representation. And once we grant that the Martian's bit-mapped memories are dispositional mental states, PP demands that we accord the same cognitive status to the "dormant" encodings in Otto's notebook.

By emphasizing the causal-functional role ("potency") of mental representations, Clark implicitly assumes that "content is as content does" (Clark, 2005a, p. 4), insofar as the semantic content of an intentional mental state is logically determined by the causal-functional role which that a state plays "at run-time" for the behavior of the cognitive agent (Prinz & Clark, 2004). According to extended functionalism, the functional role of extended mental states is physically realized by the right kind of causal interactions between an organism and its environment. Taken together, Clark's theory of mental content thus corresponds to a "wide" version of functional role semantics (Harman, 1987). This is in some sense a version of content-*internalism* (Chapter 1.1), if we count Otto-and-his-notebook as the system whose "internal" organization here-and-now determines the functional role of, and thus the semantic content of Otto's extended mental representations.

Adams and Aizawa, on the other hand, argue mostly from the perspective of externalist theories of meaning as they have been developed by Dretske, Fodor, and Millikan, although without hitching their wagon with any particular account.[27] They are, however, firmly committed to the claim that "the contents of mental representations [are analyzed] in terms of conditions on causal and/or historical relations" (2001, p. 48), which need *not* always make a difference to an agent's behavioral dispositions here-and-now. For Adams and Aizawa, the conditions which determine the intrinsic content of mental representations are therefore *not* just a matter of their (narrow or wide) functional role. Unless Clark can show that those versions of content-externalism are implausible on independent grounds,[28] it remains open for Adams and Aizawa to claim that Clark begs the question against their criticism (Adams & Aizawa, 2010). To meet their objection heads-on, I shall later take up the issue of non-derived mental content from an externalist perspective (Chapter 5). Based on a hypothesis about the "division of semantic labor" in extended cognitive processes, I will show that content externalism equally supports the claim that some external representations can have non-derived mental content.

But Clark (2005a, pp. 5-6) has raised another – and ultimately more decisive – consideration to silence Adams and Aizawa's criticism that does not hinge on

27 This interpretation is somewhat complicated by the fact that they include Searle's (1980) decidedly internalist theory of mental content in this list. But since they elsewhere clearly dissociate themselves from Searle's criterion that the content of genuine intentional states must be consciously available at least in principle (contrary to Searle, 1992), I shall simply bracket their reference to Searle.

28 For an argument against Fodor's theory of content that is independent of the HEC, see Prinz & Clark (2004).

any particular theory of intentional content. Suppose that we grant Adams and Aizawa their favorite theory of intrinsic intentionality, and also grant that a truly cognitive system as a *whole* must traffic in mental representations which have their content intrinsically. It still would not follow that each *part* of that cognitive system must consist entirely of representational states which have non-derived content. Saying that it does is to commit a fallacy of division. As Clark (ibid.) notes, Adams and Aizawa appear to concede this point, but fail to recognize its full implications. In their discussion of non-representational elements (e.g., punctuation marks, parentheses) in a hypothesized language of thought (Fodor, 1975), Adams and Aizawa admit that it is "unclear to what extent each cognitive state of each cognitive process must involve non-derived content" (Adams & Aizawa, 2001, p. 50). But if this is true of cognitive processes inside the head, we should extend the same courtesy to the beyond-the-head parts of an extended cognitive system. Because it is certainly not part of the HEC to claim that Otto's notebook considered in isolation constitutes an autonomous cognitive system in its own right.

3. Embodiment, dynamicism, and extended cognition

3.1. Cranium command or jazz improvisation?

Buzzy is a bumbling little soldier who has just been assigned the most daunting task of his life. As a fresh recruit of the Cranium Command troops, he has to pilot the brain of a twelve-year old boy, and his task is to keep the boy safe for an entire day. The main protagonists in his little army of bodily agents are Logical Left Brain, Wacky Right Brain, Hungry Stomach, Panicky Bladder, Adrenal Gland (who has a tendency to overreact), and the heart's Left and Right Ventricle. Also starring is Hypothalamus, a supporting character whose role is to regulate autonomic bodily functions. Barely out of basic training, Buzzy struggles to maneuver the boy through his day: making the boy get up and skip breakfast, running to school where the boy meets a girl, who he has to defend from bullies, being sent to the principal's office, and getting his first kiss. And that was a typical day! No wonder it's so tough becoming a successful Cranium Commander.

This is a précis of the show *Cranium Command*, a former attraction of the *Wonders of Life* pavilions in Walt Disney's Epcot Center.[29] The show purports to present the audience with a light-hearted but informative play about how the brain controls human behavior. But it is more than a play. Cranium Command has a definite affinity to the kind of *Cartesian materialism* that has been branded by Daniel Dennett (1991a) as "the view you arrive at when you discard Descartes' dualism but fail to discard the imagery of a central (but material) Theater where it 'all comes together'" (ibid, p. 107). According to this view, there is a centralized repository in the brain where the contents of our conscious thoughts and experiences are merged and effectively "enter the mind." Such a view jibes with the folk image of the conscious self as a *kybernetes* – i.e., a governor, a pilot, a helmsman who steers mind and body with reason, and acts as the central executive manager of our beliefs, intentions, and decisions.

But do not fear; I am not going to talk about such lofty things as consciousness or the seat of the self. My down-to-earth topic is how mobile creatures manage to control the movement of their bodies through space – by walking, running, swimming, or flying, and while sensing and perhaps manipulating objects in their environment. The accomplishment of those behaviors is by no means a trivial cognitive task. In fact, recent accounts of evolutionary biologists suggest that the evolution of the nervous system ("cephalization") was driven by

29 See the Wikipedia entry: http://en.wikipedia.org/wiki/Cranium_Command.

the hard problem of controlling the movements of a body that's composed of more than just a few cells – a problem significantly exacerbated with the shift towards mobile-predator lifestyles that occurred in the early Cambrian (Nielsen, 2001; Northcutt, 2002; Paulin, 2005). It is thus hardly surprising that ethologists, neurobiologists, cognitive and behavioral neuroscientists, roboticists, cyberneticists, and control theorists have devoted considerable efforts towards understanding the problem of motor control in its many dimensions. Prima facie, none of these scientific pursuits appears to be committed to any particular philosophical doctrine about the mind. And yet I claim that even in those fields, the grip of the "cranium command"-view can be felt in how the problem of motor control is typically approached. For instance, when cognitive and behavioral neuroscientists talk about motor control, they really talk about the neural substrates of motor control. When roboticists talk about motor control, they really talk about some centrally executed piece of software which controls the motors that power the robotic limbs. Let me try to illustrate the intuitive grip of the "cranium command"-view of motor control with a caricature that involves three main players – BRAIN, BODY, and WORLD, and portrays them as wrestling over who's in charge of an agent's behavior.[30]

Suppose we frame the problem of motor control as the task of guiding the movements of BODY through WORLD in accordance with certain goals and commands set by BRAIN. How could our trio make this work? BRAIN and WORLD appear to be the two main players with the power to influence BODY. However, they do not seem to be on equal footing. Whereas BRAIN seems to have direct control over BODY's movements, WORLD – as the space in which BODY has to be moved – is only indirectly implicated as the source of external stimuli for BRAIN. Therefore, states of WORLD can guide movements by BODY only in virtue of being sensed, processed, and evaluated by BRAIN. On the other hand, BRAIN can directly cause BODY to move by issuing motor commands that govern its actuators. Since the cognitively relevant effects of WORLD on BODY are always causally mediated by BRAIN, it follows that BRAIN is the real locus of control over BODY.

The "cranium command" view of the relationship between BRAIN, BODY, and WORLD has direct implications for the HEC, because the coordination and

30 The "cranium command" view bears a strong resemblance to a certain position in the philosophy of mind known as methodological individualism or internalism. Influential arguments for methodological individualism have been made by Fodor (1980, 1987, 1991), Stich (1983), Segal (1989, 1991, 2000), Devitt (1990, 1991), Egan (1992), and more recently, as explicit criticisms of the HEC, by Grush (2003), Adams & Aizawa (2008), and Rupert (2009).

control of goal-directed bodily movements is widely considered as a crucial function of cognition. If BODY and WORLD are never "actively" in charge of controlling the behavior of BODY, but merely "passive" recipients of commands issued by BRAIN, this would seem to undermine the functional equivalence ("parity") between neural and extra-neural resources vis-à-vis the task of motor control (Chapter 2). This might be taken as evidence that extra-neural resources are not part of the physically constitutive mechanisms of cognition. For instance, Kyle Butler (1998) asserts in his critique of psychological externalism that

> "[...] there can be no question that the locus of computational and cognitive control resides inside the head of the subject [and involves] internal processes in a way quite distinct from the way external processes are involved" (p. 205).

Not everybody agrees with this negative assessment, though. Growing in prominence is a renegade movement within embodied robotics (Braitenberg, 1984; Brooks, 1999; Pfeifer & Scheier, 1999; Pfeifer, Bongard, & Grand, 2007), neuroethology (MacIver, 2009), and dynamicist approaches to cognition (Thelen & Smith, 1994; Chemero, 2009; Beer, in press) who reject the strongly internalist undercurrent of "cranium command" approaches to motor control. To convey the flavor of their renegade ideas, let us revisit a classic paper by Chiel and Beer (1997) in which they argue that an exclusive focus on the internal states of the brain/nervous system seriously limits our understanding of motor control in particular and adaptive behavior in general.[31] According to Chiel and Beer, a more complete understanding of adaptive behavior requires that we shift our primary explanatory focus away from the analysis of individual components (such as the brain/nervous system) toward the causal interactions between the nervous system, the body, and the structure of the environment. Chiel and Beer state their case in a step-wise fashion. They begin by characterizing the scope of thoroughly internalist approaches to behavioral control, and then successively offer reasons for expanding one's explanatory focus – layer by layer – beyond the boundaries of the nervous system, the body, all the way into the environment (Figure 4). To characterize their own, maximally encompassing view, Chiel and Beer commend an alternative methodology for studying the cognitively relevant interactions between brain, body, and environment – an integrative-synthetic ap-

31 In their paper, Chiel and Beer (1997, p. 553) use the term "adaptive behavior" to refer to any form of behavior that enhances the survival and reproduction of an animal. In contrast, I will mostly speak of "intelligent behavior," which of course is mostly adaptive, but does not carry any purely conceptual implications about its fitness-enhancing value. We will get to Beer's (1996, 2003) more recent notion of "minimally cognitive behavior" shortly (Chapter 3.3).

proach which they call *computational neuroethology*. Using the concepts introduced in Chapter 1, we can say that their approach falls into the externalist camp. In particular, it differs from traditional internalist approaches to behavioral control in that "biomechanics and ecology are on equal footing with neural activity" (ibid., p. 556). To overcome the fixation on the "cranium command" view of behavioral control, they suggest viewing the brain/nervous system as "one of a group of players engaged in jazz improvisation," in which "the final result emerges from the continued give and take between them" (ibid., p. 555).

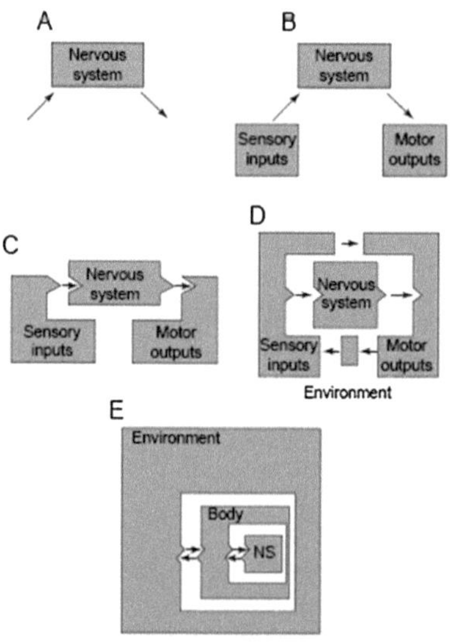

Figure 4 (from Chiel & Beer, 1997, p. 554). A continuum of increasingly "wider" explanatory strategies for studying the control of adaptive behavior (A: a thoroughly internalist perspective, E: a thoroughly embodied/embedded perspective).

How does the choice between "cranium command" and "jazz improvisation" views of behavior influence the types of questions scientists ask, and the type of methods which they use for answering those questions? As a thumbnail preview of the themes that shall concern us in this chapter, it will be instructive to review the reasons cited by Chiel and Beer for adopting their externalist point of view (Figure 4). In the above diagram, Figure (4A) represents a thoroughly internalist perspective. It encourages studies which seek to isolate specific parts of the nervous system that are causally implicated in the control of motor behavior, and then selectively analyze the properties of those neural circuits down to the level of the biophysical and molecular properties of individual nerve cells.

What – according to Chiel and Beer – is left out by adopting this perspective? First, many sensory inputs are extensively pre-processed by the body itself, and the outputs of motor neurons are significantly transformed by muscle and biomechanical properties of the body (Figure 4B). Second, changes in behavior have co-evolved over long periods of time with coordinated changes in both the nervous system and bodily morphologies, and also develop in close interaction during the ontogeny of an organism. As a result, the properties of the nervous system are extensively matched to the properties of the body, both exhibiting many complementary features. This creates both constraints and opportunities for neural control. On one hand, the nervous system evidently cannot process information that it does not possess, and its access to information is constrained by the physically possible movements which the body can perform. But more importantly, the co-evolved mechanical properties of our bodily morphology greatly simplify the problems of neural control (Figure 4C). Moreover, since organisms continuously receive and respond to feedback provided through the body, the dynamical properties of the environment play an equally crucial role for the production of intelligent behavior. It has been shown that in the absence of continuous feedback from the world, the nervous system cannot generate behaviorally meaningful patterns of activity. For instance, if an organism is artificially removed from its normal environment, its self-generated patterns of neural activity cause inappropriate forms of movement (Figure 4D). A complete understanding of how intelligent behavior emerges from the complex, reciprocal interactions among neural, bodily, and environmental factors thus requires that we study the coupled dynamics of entire brain-body-environment systems (Figure 4E).

"Cranium command" or "jazz improvisation" – these two monikers epitomize opposing poles on a continuum of views about the relative contributions of brain, body, and environment for the generation of intelligent behavior. My goal in this chapter is to motivate externalist perspectives on the problem of behavioral (esp. motor) control. On a first glance, taking Chiel and Beer's "wider" perspective seems to make the problem of motor control harder rather than easier. By attending not only to the brain, but to the whole body and its environment, including all possible interactions between those three subsystems, we introduce a large number of potentially relevant parameters that would seem to add a forbidding complexity to the problem at hand. So what analytical or empirical insights do scientists purchase by subscribing to more encompassing views of motor control? Finally, and most importantly, do these considerations support an argument for the HEC?

3.2. Information-theoretic implications of embodiment

3.2.1. Hexapod walking

Hexapod creatures – creatures that walk on six legs such as stick insects or cockroaches – offer intriguing possibilities for studying walking behavior (Cruse, Dean & Ritter, 1998; Dürr et al., 2003). First, evolution has already populated pretty much every corner of the earth with six-legged insects, so we have massive evidence that hexapod walking is a highly adaptive type of locomotion. Using six legs for walking simplifies the tasks of balancing and stabilizing the body in comparison to tetrapodal or bipedal walking. This allows for greater behavioral flexibility as well: since not all six legs are required (at any given moment) to maintain a stable position, the other legs are freed up for other tasks (e.g., reaching for new foot placements, or manipulating objects in the environment). Moreover, having six legs adds fault-tolerance to the walking system: if one of an insect's legs is disabled, its walking capacity remains largely intact. Second, since insects have a much simpler nervous system than most other animals, it is possible to pin down the exact organization of the neural circuitry that mediates between sensation and motor behavior. Finally, the body of stick insects has a very research-friendly morphology. Their legs stick out to the side, unoccluded by other body parts, and it is relatively easy to strap recording devices to the insect's straight torso to monitor its walking. It is because of all these features that biologists, ethologists, bio-cyberneticists, and neurophysiologists have gained a fairly detailed understanding of the control mechanisms that underlie hexapod locomotion. Many hexapod robots have been designed on the basis of biologically realistic theories of insect walking; at the same time, biologists have frequently used hexapod robots as fertile testbeds of their hypotheses about the coordination of insect walking. To introduce some basic design principles of embodied motor control, let us take a look at the groundbreaking research by Holk Cruse and his colleagues on hexapod walking, gained from his long-term study of stick insects (e.g., Cruse et al. 1998; 2002).

In the course of walking, a single leg always performs two kinds of movements: a power stroke ("stance phase") and a return stroke ("swing phase"). During the stance phase, the legs support the weight of the body while pushing it forward. During the swing phase, the leg is lifted off the ground, moved forward relative to the body, and put down again to initiate another stance phase. The dynamics of hexapod walking shows several typical gait patterns. When there is little load to carry and the animal is at liberty to walk as fast as possible, hexapods usually fall into an alternating tripod gait. This means that there are always three legs touching the ground at a given moment in time: the front and hind leg of one side, and the middle leg of the other side. Once the animal is forced to slow down (e.g., because of carrying an increased load), it automatically shifts to a tetrapod gait where at least four legs are always simultaneously on the ground. In this gait, the order in which the legs perform return strokes on both sides of the torso generates a characteristic wave-like pattern. Occasionally, the animal crawls, moving just one leg at a time.

What is the mechanism which controls the walking behavior of hexapods, and the transitions between the different gaits? In principle, two different kinds of control architectures could be built to coordinate insect walking: centralized and decentralized control architectures. A centralized controller (e.g., a central pattern generator located within the insect's brain) would directly specify the relevant state transitions for all six legs. But as the research by Cruse et al. shows, stick insects rely on a decentralized architecture consisting of six more or less independently working controllers (i.e. one for each leg) that are connected in a parallel network, and whose global coordination is possible only through their local interactions with the environment (Cruse et al., 1998).

As a first step of his research, Cruse derived a set of six basic rules that reflect the decentralized motor coordination between the suggested six subsystems of walking (L1-L3, R1-R3; see Figure 5). Each subsystem consists of a neural network controlling the behavior of a single leg. As an example, consider some of the rules which describe how the behavior of leg L1 is influenced by its two nearest neighbors, legs R1 and L2. If L2 is in return stroke, the start of the return stroke of L1 is inhibited (Rule 1). The start of a power stroke in L2 or in R1 excite the start of a return stroke in L1 (Rule 2). When R1 or L1 is in caudal position, they excite in each other the start of a return stroke (Rule 3). There is no central controller through which the individual subsystems communicate.

The core of Cruse et al.'s model of hexapod walking is WalkNet, a parallel assembly of artificial neural networks each controlling the behavior of a single leg. The outputs of WalkNet specify the three angles (α, β, and γ) of the joints which determine the position of a leg (Figure 5, left). A WalkNet for a single leg consists of several subnets, the most important of which are the *stance net*, the *swing net*, and the *selector net*. The swing net has the relatively easy task of performing the return stroke. The stance net, which controls the power stroke, has to juggle a variety of different tasks, including curve walking, velocity control, or reacting to changing loads. Importantly, it also has to allow that the animal's legs become properly coordinated so as to prevent the occurrence of lateral forces in the torso.

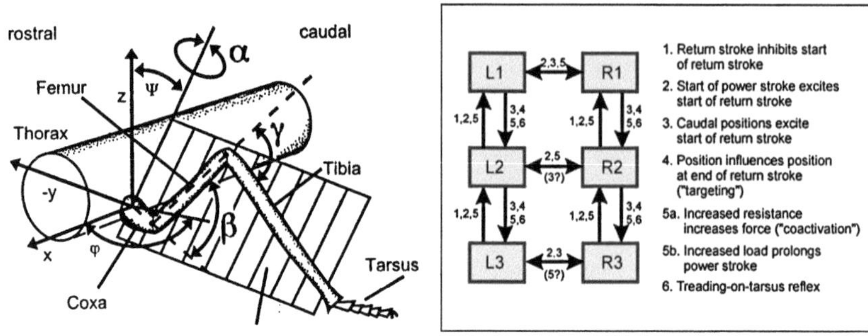

Figure 5 (from Cruse et al., 1998). Left: leg geometry of a stick insect. Right: A basic rule set for describing stick insect walking. The arrows show in which direction the rules are applied.

As Cruse was able to show, the coordination between the six subsystems is not centrally managed by neural control structures, but by actively exploiting the interactions between the insect's body and its environment. As an illustration, consider a situation in which the insect stands on the ground, and pushes one of its legs backwards to perform a power stroke; the other legs standing still. This pushes the body forward. But in virtue of the simple fact that all legs are physically attached to the torso, it also instantaneously changes the joint angles of the other legs that are planted on the ground. Importantly, it specifically changes the angles in the right direction – since the body is moving forward, the other legs will automatically be pushed forward, and the joints will be bent or stretched accordingly. The directional information generated by those changes is thus provided "for free" by the morphology of the insect's body. The resulting changes in the position of the legs can be detected by angle sensors that are attached to the insect's legs, and connected to the neural leg controllers. As this example shows, the global coordination of the legs is enabled by a feedback loop which runs through the agent's body and its environment.

Finally, the job of the selector net is to control which of the two main subnets (stance or swing net) gets to control the action. This is necessary because the two subnets are never turned on or off, but always remain active – the selector net simply gates out one of the two, allowing the other net to gain control over the actuator. The selector net thus acts as a kind of neural oscillator, but one that is effectively driven by the environment. This is because inside the selector, the nodes which are responsible for initiating either the power stroke or the return stroke use a positive feedback mechanism. Hence without constant input from the environment, whatever net is currently active would become activated even more, and the controller would get caught in a self-reinforcing loop.

Over the past two decades, Cruse's WalkNet platform has become a well-researched framework for studying hexapod walking. Refined versions of its architecture have been implemented on a variety of hexapod robots which perform increasingly well in ecologically realistic environments (Dürr, Schmitz, & Cruse, 2004).

3.2.2. Three principles of embodied control design

The decentralized architecture of WalkNet illustrates some key design principles of "embodied" control strategies, as outlined in recent work by Rolf Pfeifer and his colleagues (Pfeier, Iida, & Bongard, 2005; Pfeifer et al., 2007, Chapter 4). The suggested design principles reflect a "deeper" sense of embodiment which goes beyond the uncontroversial fact that physical agents have a body. Surely this trivial sense of embodiment has not been ignored by mainstream roboticists either, e.g., when they are trying to deal with physical factors like gravity, friction, torque, inertia, energy dissipation, etc. What they often did ignore, however, is that some aspects of the control task itself can be taken over by the interplay between the material structure of an agent's body, the geometry of its morphology, and the structure of the environment. The goal of this section is to bring what we may call *information-theoretic* implications of embodiment into closer view.

Before we proceed, let me clarify what is meant by "design principles." Coming from an engineering perspective, Pfeifer et al. (2007) are primarily concerned with *building* stuff. Hence one of their main goals is obviously to provide the emerging field of "embodied" robotics with a principled set of strategies for building adaptive, autonomous intelligent systems capable of coping with real-world environments. But at the same time, since their design of embodied model agents is directly based on their theoretical understanding of embodied cognition, those principles are equally meant to express some fundamental principles of naturally occuring intelligence (Pfeifer et al., 2007, p. 356). Their attempted symbiosis of science and engineering stems from a commitment to the synthetic methodology of "understanding by building" (cf. Braitenberg, 1984).

Loose couplings

An important principle of embodied agent design is the importance of *parallel, loosely coupled processes* for the success of decentralized control schemes (Pfeifer et al., 2007, Chapter 4.9). The term "loosely coupled" is used here in contrast to hierarchically coupled sequential processes found, e.g., in conventional computer programs, such as when a program calling a subroutine has to wait for the subroutine to execute its task before it can continue. Such hierarchical control schemes are examples of strongly coupled processes. In contrast, the coordination between the six single-legged control systems of Cruse's WalkNet is achieved by a large number of parallel, loosely coupled processes that run asynchronously, and in ways that are directly tied to the agent's sensorimotor apparatus, and the dynamics of its interaction with the environment. While loose couplings are an important design feature of embodied intelligence, we should note that embodiment also provides for extremely strong forms of coupling, such as when two joints are linked by a direct physical connection (e.g., a bone). In Cruse's WalkNet, for example, we saw a nice example of how strong and loose couplings interact in the service of coordinating the insect's legs: the anatomical fact that all legs of an insect are connected to a single torso (= strong coupling) was exploited in tandem with the mechanical interactions between the legs and their local environment (= loose coupling).

In this context, it is worth drawing attention to the "bone-brain continuum" hypothesis – a neuroethological conjecture about the ways in which the load of motor control is generally distributed across brain and body (MacIver, 2009, pp. 491f). The basic idea is that regularities in agent-environment interactions which occur over disparate timescales are reflected in the design of different types of control systems. For instance, among the longest timescale regularities for mobile organisms is the need for a balance of forces along their preferred axis of travel, which is moving in straight lines. This long-term regularity has been biologically "absorbed" at the level of structural tissue (bones, cartilage, chitin), most evidently by the predominance of bilaterally symmetrical body plans. Since the maintenance costs of those tissues are the lowest of any body material, strong couplings are a very cost-effective control strategy in terms of energy use. In contrast, short time-scale regularities, such as the current state of self-motion, can only be partly absorbed at the structural level (e.g., by strategically positioning sensors in morphologically well-chosen spots), but generally involve much more processing at the neural level. Neural processing is a much more expensive strategy of sensorimotor coordination, because maintaining neural tissue requires forty times more power per unit mass than bone. The "brain-bone continuum" hypothesis asserts that the distribution of control processess which an agent needs to perform in order to move smoothly through space evolves in ways that roughly correspond to the suggested cost-benefit ratio. From this perspective, "there is no basis for an invidious distinction between bone and brain" (ibid, p. 492).

Designing for emergence

The design of adaptive control strategies for embodied agents should aim to incorporate the effects of emergence (Pfeifer et al., 2007, Chapter 3.6). This sounds a bit like a contradiction if we take the notoriously ambiguous term *emergence* to describe behavior that is somehow unpredictable or underivable from a set of prior assumptions. But the term *emergence* is used here to refer to certain aspects of an agent's behavior which are not the result of operations that were pre-programmed into the agent's on-board control apparatus. Instead, they arise as a "side effect" of a carefully balanced cascade of interactions between an agent and its environment. Clark (2001a, p. 114) has dubbed this notion of emergence as a kind of *unprogrammed functionality*. For instance, the inter-leg coordination achieved by Cruse's WalkNet is an emergent effect of the interactions between the neural controllers, the morphological and mechanical properties of the legs and the torso, and the ground on which the insect is crawling.

The contrast between preprogrammed and emergent behavior is nicely exemplified by Steels (1994), who compares a centralized and a decentralized control strategy for getting a robot to position itself between two poles (so it can recharge its batteries). A centralized and heavily pre-programmed strategy would be to equip the robot with sensors that measure its current position relative to the two poles, encode their inputs, and then pass them on to a subroutine which computes the most effective path to the prespecified target position. In contrast, Steels describes an alternative, behavior-based strategy that factors in the reliable effects of a series of agent-environment interactions. For the strategy to work, a light source is used to indicate the position of the charging station. The target behavior of Steels' robot emerges from the interactions between two parallel, loosely coupled modules: a phototaxis-driven system that leads the robot towards the light source, and an obstacle-avoidance system that causes the robot to retreat and turn away when it hits something. The "side effect" of the local interactions between the relatively simple computations performed by the robot's on-board machinery, the morphology of its body, and the structure of its environment is that the robot will gradually drift towards the light source. If the robot collides with one of the poles, it retreats and approaches the light source from a new angle. Eventually, it comes to a halt in the only position in which the two behavioral systems are in equilibrium – close to the light, but without hitting either pole – and recharges.

Steels himself characterizes the pole-oriented behavior of his robot as an example of an emergent, "uncontrolled" property, but this terminology is somewhat misleading (cf. Clark, 1997, pp. 108-109). The behavior is "uncontrolled" only insofar as we only look at the agent's internal mechanisms – i.e., in this case, the software that controls the two behavioral modules. But such an internalist perspective is itself part of the "cranium command" view of motor control which embodied roboticists like Steels seek to overcome. Viewed through an externalist lens which takes into account the reliable effects of iterated interactions between brain, body, and environment, the observed patterns of the robot's behavior are neither uncontrolled nor even unexpected. In the artificial environment of the lab, they are the predictable result of canny methods of embodied design; in nature's lab, the efficient exploitation of agent-environment interactions is often the adaptive result of learning or evolution.

An important implication of emergence is that there may be a considerable gap between the representational content of the sparse neural control signal and the indirect behavioral effects to which it leads. For instance, we have seen how the locomotory movements of Steels' second robot are driven by two largely independent modules. One module fires in response to the presence of light, while the other module fires in response to the presence of macroscopic objects in its surroundings. In light of our knowledge of how those modules have been designed, the direct function of the first module is to detect and move towards light (*phototaxis*), while the direct function of the second module is to avoid obstacles. Accordingly, it makes sense to describe the representational content of the outputs delivered by the motor controller as commands of the form "Move closer to light!" or "Avoid pole!"[32] Clearly, the direct function of these motor signals is different from the indirect behavioral behavioral effects (such as the robot docking into the charging station). This implies two things. First, if we wouldn't know exactly how the two robots are configured internally, we might be misled into thinking that the second robot is driven by the same sophisticated neural control architecture as the first robot. But this would clearly amount to an "overattribution fallacy" (Hutchins, 1995a, p. 356; see Chapter 5.1). Second, it also shows that the suggested notion of emergence is a matter of degree. The larger the representational gap between the actual motor signal and the observed target behavior, the more we are inclined to call the target behavior an emergent phenomenon (in this sense).

The Principle of Ecological Balance

A third key design principle exemplified by Cruse's WalkNet is the *Principle of Ecological Balance* (Pfeifer et al., 2007, Chapter 4.8). It refers to the ways in which the task of controlling some aspect of an agent's behavior can be spread *in certain surprising, and often unintuitive ways* across brain, body, and environment.[33] More specifically, the principle asserts a certain trade-off as the con-

---

32 I am glossing over some tricky conceptual questions here about how to individuate the representational content of mental states; for further discussion, see e.g., Dretske (1986, 1991), Fodor (1990), and Millikan (1993).

33 Clark (2008, p. 7) subsumes the "Principle of Ecological Balance" under the phenomenon of *nontrivial causal spread* (cf. Clark, 1998b; Wheeler & Clark, 1999; Wheeler, 2005).

sequence of a balanced task distribution: if an embodied agent effectively exploits the material, morphological, and dynamical properties of its body, it can thereby greatly ease the complexity of the control task which the neural system has to perform (while maintaining the same level of adaptive success). Since lessening the computational load on the neural controller leads to a reduction of overall metabolic costs, an ecologically balanced task distribution supports more energy-efficient ways of maintaining the target behavior (cf. Clark, 2008, Chapter 1). Let's highlight the Principle of Ecological Balance with a few concrete examples.

In the domain of sensory processing, a striking demonstration of ecological balance enables the female (robot) cricket to identify and locate male conspecifics by their song, and then move toward them (Webb, 1994; cf. Clark, 2005b). A cricket has two ears, each located at its forelegs (the left and right *tympanon*), and connected through an internal tracheal tube. The tube also connects the insect's ears to two other openings (*spiracles*) located at the top of its body. Because of the tube, sound waves created by environmental events travel to each ear along two different routes: a direct external route (from the external source) and an indirect internal route (via the other ear, the spiracles, and the tracheal tube). At the ear which is closer to the external source, the sound which arrives through the direct "outer" route will have travelled less far than the sound which arrives though the indirect "inner" route. Because of the implied temporal delay, the two signals will be out of phase – more so for the ear closer to the source than for the ear farther from the source. At both eardrums, the out-of-phase waves are summed up, which causes a vibration of greater amplitude (i.e., a louder sound is heard) that what could otherwise be detected by the cricket's relatively simple neural circuitry. Because of the way in which the cricket's body is built, the interneuron which is connected to the ear that is closest to the sound source will fire first; other parts of the nervous system are set up to turn the cricket into the corresponding direction. Since the male cricket's song is broadcast in discrete outbursts, the female cricket can continually adjust the direction in which it moves at the start of each outburst.

If we only inspect the neural circuitry of the female cricket, we would find that it does not possess the special kind of representational capacities (e.g., neural mechanisms for the recognition of particular song patterns, sound localization, movement planning) that would seem to be required to recognize her mates. But that is only part of the whole story. The cricket manages to get by with its sparse neural control structures because it actively exploits a fine-tuned complementarity between task-relevant neural, biomechanical, temporal, and environmental factors. First, the fixed-length tracheal tube plays an indispensable role in filtering, processing, and amplifying auditory inputs. It transmits only sound of the desired calling song frequency, and the phase shifts which this transmission generates are geared to that particular frequency band. This allows the female cricket to identify the song of her male conspecifics, and to localize the source of the perceived sound. Second, the adaptive success of the cricket's behavior also depends critically on the discontinuity and repetition rate of the male cricket's song.

Bat ears and fly eyes are other intriguing examples of ecologically balanced interactions between bodily morphologies and the demands of sensory processing (cf. MacIver, 2009 for disucssion). For the bat as an active-sensing animal, the intricate shapes of the bat's outer ear (*pinna*) and the small flap of tissue coming up from the base of the pinna (*tragus*) provide spectral cues to the vertical elevation of a target. Similarly, the non-homogeneous arrangement of facets in the fly's eye compensates for the phenomenon of motion parallax, which vastly simplifies the computation of optic-flow patterns that is necessary for the fly to stabilize its flight.

Another proof-of-concept of the Principle of Ecological Balance is the running robot "Puppy," which is capable of controling fast locomotion by means of surprisingly "cheap" design strategies (Iida & Pfeifer, 2004; see also Clark, 2008, Chapter 1). Rapid locomotion poses considerable challenges for motor control, because it cuts down on the time that is necessary for feedback-controlled mechanisms (which are normally used in walking robots) to receive sensory inputs. The designers of Puppy responded to this challenge by building a robotic creature which is capable of maintaining a robust running gait without using any sensory feedback, while relying purely on the self-stabilizing dynamics that are commonly found in biomechanical systems. The four-legged design of Puppy mimics some basic properties of a muscle-tendon system (Figure 6). Each leg consists of a lower and an upper part, connected by a spingly passive joint. Forelegs and hindlegs are connected by a flexible spine. The only control signals used are synchronized oscillations that are applied to the motors attached to Puppy's hip and shoulder. Those signals do not specify any gait-related properties such as the difference between stance and flight phase, speed and acceleration, or the angular positions of the joints.

Despite the extremely sparse nature of the control signal, and the complete lack of sensory feedback, we can see from Figure 6d that Puppy's running gait settles into a stable periodic cycle between a stance and a flight phase (i.e., an interval, characteristic of running, during which no feet are on the ground). The dynamics of a natural quadrupedal gait emerged from the interplay between the agent's morphology, a simple control signal (the amplitude and frequency of oscillation), intrinsic material dynamics (as determined by the spring system of the legs, the flexible, elastic spine, and the proper mass distribution), and the physical structure of the environment (gravity, friction, shape of the ground). The self-stabilizing property of this complex dynamical system demonstrates the adaptive effects of an ecologically balanced task distribution. As it turned out, even the weight of the battery, which was attached to the spine, seemingly played an instrumental role in distributing the mass properly. Moreover, the authors note that the stabilization worked only if the foot-ground contact provided little friction. In the experiment, the necessary amount of slippage was guaranteed by the fact that Puppy's legs were made of bare aluminum. If Puppy was made to wear rubber boots for a better grip, it was in fact more likely to trip. Thanks to the self-stabilizing dynamics of Puppy's walking behavior, it was possible to control the speed of its forward motion by simply varying a single phase parameter of the oscillators, namely the temporal delay between the foreleg and hindleg motors. The design of Puppy's embodied and embedded scheme of motor control is a nifty demonstration of what we have called "designing for emergence."

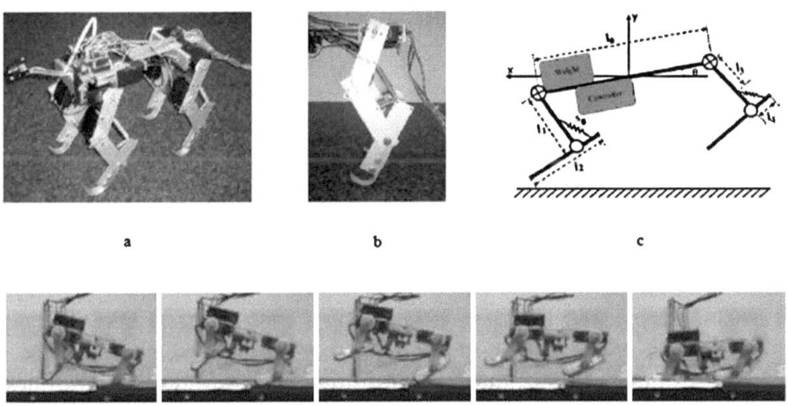

Figure 6 (from Pfeifer, Iida, & Gomez, 2006). (a) Picture of the quadruped running robot "Puppy." (b) The spring system in the hindlegs. (c) Schematic design of Puppy. (d) Puppy running on a treadmill.

Our third and final example of ecological balance concerns the sensorimotor coordination of grasping behavior. In order to grasp something, the grasping hand must be brought into a position where the fingers can be firmly wrapped around the object. Most robotic hands in use today are built from completely rigid materials. They are controlled by sophisticated algorithms that require exact representations of the shape of the arm, the fingers, and the to-be-grasped object, and rely on numerically precise feedback signals provided by pressure, angle, and torque sensors to adjust the fingers closer to the desired goal state. In contrast, a human hand is made from partly non-rigid materials such as elastic tendons, deformable finger tips, and deformable tissue between the fingers. The easily-adjustable mechanical properties of muscles, tendons, ligaments, and bones all provide a kind of "passive compliance" that significantly simplifies the neural coordination of grasping movements. The embodied design of the "Yokoi Hand" (Yokoi et al., 2004), which can be used as a robotic or a prosthetic device, illustrates the Principle of Ecological Balance between materials, morphology, and neural processing. Thanks to its anthropomorphic design, the basic version of the Yokoi-hand operates with a "cheap" control scheme (called a "CLOSE") that does not have to "know" in advance the exact shape of the to-be-grasped object, and does not require any exact feedback signals. One and the same application of a "CLOSE" can support the successful grasp of differently shaped objects, because part of the shape adaptation is performed by the elastic, flexible, and deformable parts of the Yokoi-hand as it touches the object.

3.2.3. Morphological computation and "control-parity"

What is the upshot of the Principle of Ecological Balance? In the previous section, we have seen examples of how the cunning exploitation of the relationship between neural processing, bodily morphology, sensor placement, materials, and environmental structure can significantly transform the problem of behavioral control, especially by simplifying the control task which the nervous system has to perform. The result of this task distribution is commonly characterized as a *trade-off* between "cheap" morphology and "expensive" control. However, this manner of speaking assumes that the extra-neural resources which are part of this equation are not strictly speaking doing any controlling. Instead, they only reduce the control load that must be shouldered by the brain. One of the reasons for this persistent dualism between morphology and control is the underlying assumption that only the nervous system performs behaviorally significant *computations*. In short, no control without computation! Let us call this view the "standard" interpretation of the Principle of Ecological Balance.

As I will argue in this section, the standard interpretation is misconceived. Perhaps inadvertently, it perpetuates the "cranium command" view by maintaining a strict dichotomy between mechanical design and control design that permeates mainstream robotics, artificial intelligence, and neurobiology. In order to challenge this dichotomy, one of Pfeifer's former students, Chandana Paul (2004), devised a nifty thought experiment to break down the theoretical boundary between morphology and computation. She describes several simple robots whose bodily morphology clearly assumes certain well-defined computational roles that are vital for controlling their own behavior. Her examples show that the causal spread of responsibility for adaptive sensorimotor behavior can properly be understood as spread of computational and control labor. Therefore, computation ain't (all) in the head! The strategy of off-loading computations from the brain into the body is known in embodied and biomimetic robotics as *morphological computation* (Pfeifer et al., 2007, p. 96).

To appreciate the basic idea of morphological computation – computations that are performed courtesy of "brute-causal" features of the body and how it interacts with the environment – let us consider Paul's simple "XOR-robot." It has only one wheel, with two actuated degrees of freedom. The actuation of motor M1, which is controlled by a simple perceptron that computes a logical OR on inputs A and B, causes a wheel to turn so that the robot moves forward. The actuation of motor M2, which is controlled by another simple perceptron that computes a logical AND on inputs A and B, causes the wheel to be lifted off the ground so that its rotations become idle (Figure 7, left).

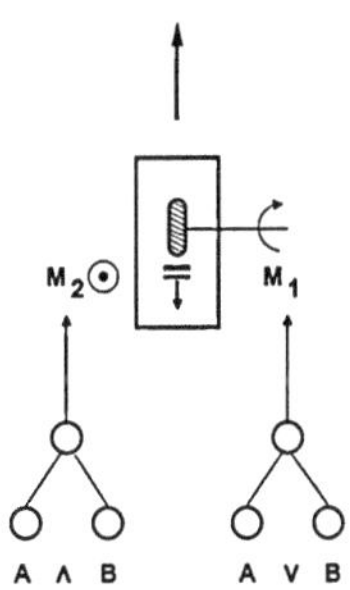

A	B	Behavior
on	on	stationary
on	off	moving
off	on	moving
off	off	stationary

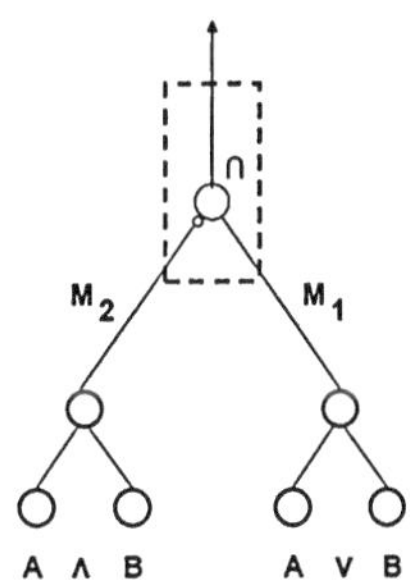

Figure 7 (from Paul, 2004). Left: Design of the XOR-robot. Middle: Formal correspondence between the behavior of the robot and an XOR-function. Right: Boolean function computed by the XOR-robot, which corresponds to the function M1 AND ~M2.

A single-layer perceptron is the simplest form of a feedforward neural network. It has the computational capacities of a linear classifier, i.e., it can learn to compute any function as long as the function is linearly separable (Rosenblatt, 1962). For instance, it can compute every function that is truth-functionally equivalent to an AND or an OR. However, it cannot compute any function that is truth-functionally equivalent to an XOR ("exclusive-or"), because XOR is a function which is not linearly separable. However, if we look at the pattern of motor outputs produced by Paul's robot, under the four possible input conditions, the robot behaves as if it computes a function that is truth-functionally equivalent to an XOR (Figure 7, middle). Hence it seems that the robot is able to systematically behave in ways that its dedicated neural control systems (i.e., the two perceptrons) cannot possibly compute. *Mirabile dictu*! How can this be?

The apparent paradoxon disappears once we expand the relevant unit of computational analysis from the level of the neural networks to the larger system which includes the robot's entire body. In Paul's example, the morphology of the embodied robot provides an additional layer of processing which applies a truth-functional NOT to the state of M2, and then combines the resulting state with the state of M1 in form of a truth-functional AND (Figure 7, right). Taken as a whole, the system thus computes an XOR.

Concerning this feat of what Paul calls "latent morphological computation," one might argue that the robot body performs computations only "in the eye of the beholder," i.e., as a matter of conventional interpretation. A less observer-dependent manifestation of morphological computation would seem to require that the robot itself actively exploits the computational functions of its bodily movements. To envision this kind of situation, Paul embeds her XOR-robot into a real-world "vacuum-cleaning" robot which receives external sensory input (via a camera), moves around, and – crucially – can also sense the behavioral effects of its own movements (via an accelerometer on its front wheel). Embedded in this concrete task environment, the robot actively generates bodily movements to perform XOR computations whose outputs are directly linked to various other control routines that govern the robot's vacuum cleaning. As a result, the control structure provided by the bodily computations is now functionally indistinguishable from the conventional control structure that is involved at different stages of processing. This shows that in the vacuum-cleaning robot, the computational labor which is associated with the control task is literally spread out across the robot's brain and its body. Consequently, morphology and control can no longer be seen as two opposing concepts. This shows that neural and morphological resources are equally capable of assuming concrete, adaptive roles within the larger information-processing economy of embodied agents.

What further philosophical lessons we can learn from Paul's compelling demonstration of embodied sensorimotor control? First, the idea of morphological computation which straddles the traditional gap between mechanical and control properties reinforces the distinction between *pragmatic* and *epistemic* actions (Kirsh & Maglio, 1994; cf. Chapter 2.1). For the role of the robot's actuation of the front wheel is not just to push the vehicle forward, which would be an instance of pragmatic action. At the same time, the body-involving XOR-loop plays a crucial epistemic role within the control architecture of the robot. It functions like a regular logic gate, implementing computations that control the robot's intelligent behavior. Second, the notion of morphological computation ties our discussion back to the thesis of extended functionalism and the *parity principle* (PP). We began this chapter by quoting Kyle Butler's (1998) critical assessment of the HEC, which was based on his claim "that the locus of computational and cognitive control resides inside the head of the subject [and involves] internal processes in a way quite distinct from the way external processes are involved" (ibid, p. 205). But our excursion into biological cybernetics and embodied robotics has given us reason to doubt the generality of Butler's claim, because we have seen cases in which the task of behavioral control was literally shared by neural and extra-neural resources.

One way to make explicit the information-theoretic implications of the Principle of Ecological Balance is by way of a cognate version of PP – one that specifically deals with the identification of *control* processes:

(CONTROL-PARITY) If, as we confront a task, a part of BODY (WORLD) functions in a process which, were it to go on in BRAIN, we would accept as part of a process of behavioral (e.g., motor) control, then that part of BODY (WORLD) *is* (for that time) part of the control process.

The principle of control-parity is what proponents of embodied and situated cognition have in mind when they call for a reconceptualization of the classical notion of *control*. For instance, Pfeifer et al. (2007, p. 362) illustrate their rejection of the "cranium command" view of control with an astute observation about the dynamics of walking downhill. When I am running on a flat terrain and all of a sudden the path starts going downhill, I automatically begin to run faster. This is not because my brain sends an instruction to my muscles to accelerate. Instead, as my body is pulled downwards by gravity, its motions are naturally sped up, which leads to an acceleration of the neural oscillators in my brain (and spinal cord). The suggested role-reversal between BRAIN and BODY and/or WORLD thwarts the internalist intuitions that BRAIN is always the sole locus of behavioral control (Chapter 3.1). As Pfeifer and Bongard (ibid.) summarize the deeper implications of embodiment, "it is not the neural oscillators alone that determine the behavior of the body, but the body just as much determines the frequency of the neural circuits. In other words, there is mutual coupling, or to use dynamical systems jargon, there is mutual entrainment." Similarly, an implicit commitment to control-parity seems to undergird Chiel and Beer's (1997) previously cited claim that "biomechanics and ecology are on equal footing with neural activity" (ibid., p. 556), and their dynamicist vision of the brain/nervous system as "one of a group of players engaged in jazz improvisation" in which intelligent behavior "emerges from the continued give and take between them" (ibid, p. 555).

Once we have reason to accept the principle of control-parity, it only takes another application of Clark and Chalmers' original principle of parity to infer the stronger HEC. That is, if it can be established that *control* processes are extended, it follows that *cognition* is extended provided that the functions associated with neural control processes are literally taken over by the canny tweaks of bodily dynamics, biomechanical constraints, morphological affordances, and epistemically potent worldly actions. The plausibility of this latter inference rests on general considerations in favor of extended functionalism that we have already presented in Chapter 2. In the remainder of this chapter, I shall further examine the principle of control-parity from the perspective of dynamical systems theory-based cognitive science. But first, I want to connect our present discussion of embodiment to the previously mentioned (Chapter 1.4) theoretical alternative between the "hypothesis of extended cognition" (HEC) and Robert Rupert's "hypothesis of embedded cognition" (HEMC).

3.2.4. From morphological computation to epistemic interactivity

In a series of papers that has culminated in a recent monograph, Robert Rupert (2004, 2009, 2010) vigorously argues that psychological externalists have overplayed their hand against orthodox psychological internalism, and that it's time to call their bluff (see also Chapter 1.4). Rupert argues that the HEC compares unfavorably to a suitably modified internalist perspective according to which cognitive processes occur entirely within the organism, but sometimes "depend very heavily, in hitherto unexpected ways, on organismically external props and devices" (2004, p. 393). The "hypothesis of embedded cognition" (HEMC) calls for a science of brain-bound (or at least organism-bound) cognitive systems which depend in various ways on the environment, but stops short of claiming that the environment is literally part of cognition. Rupert's "gambit" is to pit the HEC against the HEMC in terms of an overall cost-benefit analysis. His basic strategy is to argue that whatever insights are generated by adopting the methodologies of embodied and situated cognition do not require the HEC, because they can all be accommodated on the basis of the simpler and more conservative HEMC. Hence cognitive science would accrue no real methodological benefits, but incur significant ontological and methodological costs if it were to prefer the HEC over the HEMC (for a discussion of those putative costs, cf. Chapter 6.2). Therefore, HEMC outperforms HEC on general grounds of either conservatism or theoretical parsimony.

Rupert's gambit can be succinctly cast as a dilemma-style *conservatism-or-simplicity* argument against the HEC (see esp. Rupert, 2009, Chap. 3.3, pp. 45-46; Chap. 4.1; Rupert, 2010, pp. 346-47):

1. Either HEC recreates the "standard" internal cognitive architecture in its models or it doesn't.
2. If it doesn't, it loses out to the HEMC on grounds of explanatory power.

3. If it does, then the HEC is not substantially (but merely terminologically) different from the HEMC, so the HEC loses on grounds of conservatism.
4. If the HEC posits additional extended cognitive systems, then the HEC loses on grounds of simplicity.

In response to this dilemma, I want to bring into view a lingering misconception of the alternative between a HEC-friendly and a HEMC-friendly methodology which is manifested in Premises (3) and (4) of Rupert's argument. As for Premise (2), it remains a safe bet that a fair amount of cognition is going on entirely inside the boundaries of the brain (and thus the organism). Because of the deeply pluralistic vision of cognitive science that HEC-theorists usually harbor, there is no reason why externalist frameworks should not be allowed to avail themselves of good "orthodox" explanations that we already have of particular brain-bound cognitive processes.[34] This doesn't mean, however, that any of the orthodox research programs have correctly identified what Rupert calls our so-called "standard" internal architecture; but we will get to that issue later (cf. Chapters 4-5 below).[35]

In support of Premise (3), Rupert (2010) compares the two competitor perspectives as follows: "According to HEMC, there exists an internal cognitive system, some external materials, and interaction between the two. According to HEC, there exists an internal system (which is cognitive), some external materials (also cognitive), and interaction between the two" (p. 10; see also 2009, p. 19). Based on this comparison, he argues that HEC advocates an unmotivated form of revisionism that merely adds the label "cognitive" to one of the three kinds of things recognized by both approaches. It is quite telling that Rupert's portrayal attaches no specific status to the nature of the interactions between internal and external resources. Are *they* cognitive or not? This omission is no coincidence. It is common among critics to think of the current debate over the bounds of cognition as primarily a dispute over whether external resources are "intrinsically" cognitive or not (see also Clark, 2007). But from a methodological perspective, the more pressing issue is rather how we should understand the cognitive significance of the complex two-way and three-way interactions between brain, body, and environment in the production of cognitive outcomes. Adding this category to our comparison yields an important contrast that is missing from Rupert's presentation. For the HEC, interactivity can *directly* assume a

34 Rupert (2009) seems to agree with this when he notes that "I have not claimed that, if the extended view is true, there must be a distinctively extended-cognition based explanation of every cognitive phenomenon" (p. 41).

35 I will discuss Premise (4) in Chapter 3.3.2 below.

cognitive role, whereas for the HEMC, the cognitive role of interactivity must always be *mediated* by the brain-bound cognitive apparatus.

As an illustration of this contrast, Kirsh (2009, pp. 268-90) has shown how an appropriate grasp of epistemic interactivity "in-the-world" can radically alter our conception of problem-solving, and our choice of methods that we deem appropriate for studying it. In the classical, artificial intelligence-based approach, it was assumed that people thoroughly search an internally represented problem space before acting, and their interactions with the environment do not reshape the nature of this problem space or the search process. Consequently, the processes of *framing* (i.e. finding the right representation of a problem space) and *registration* (the activity of mapping that representation onto the world) were merely seen as ways of preparing for the process of problem-solving. They were never viewed as actual means of solving it, which was essentially reduced to the problem of internal search. Contrary to what Rupert's construal of the debate might suggest, a proper understanding of situated problem-solving does not just—or even primarily—consist in paying closer attention to the cognitive role of external representations. For instance, the modified classical theory (Larkin & Simon, 1987; see also Vera & Simon, 1993) was well aware of the fact that problem states can be encoded internally, but also partially encoded externally. But for Larken and Simon, the perceived value of external representations was merely to function as analogs of internal search, designed to make the same search process faster and more efficient. This concession accommodated some interesting facts, e.g., that the surface representation of a problem strongly determines the problem-solving trajectory (Zhang & Norman, 1994). However, it did not lead to a more radical shift in focus and methodology that is necessary to capture the most potent aspects of epistemic interactivity in-the-world.

For the situationist, finding the right problem-space is seen as the most critical part of solving a problem, because it determines to a great extent the nature of the problem that people are actually trying to solve. From this perspective, search itself is interactive, and highly entwined with iterated loops of framing, representing, and registering a given task domain. As an example, consider this clever method of measuring the height of trees, reportedly used by some Native American tribes (Surendran, 1996). A tribesman would first seek out a spot from which he could just see the top of the tree, looking at it upside down between his legs (Figure 8). Assuming this position creates a task-specific frame of reference in the world – rather than in one's head – in which the easily walk-able distance to the base of the tree directly corresponds to its height, since the angle of the resultant line of sight is about 45° for a normal adult. From a methodological standpoint, reserving the label "cognitive" strictly for the problem-solving contributions of the brain is no more helpful than just extending the label "cognitive" to head and eyes, bodily posture, stepping behavior, or the ground. It is their carefully orchestrated interplay and functional integration *in situ* which deserves most of our cognitive praise.

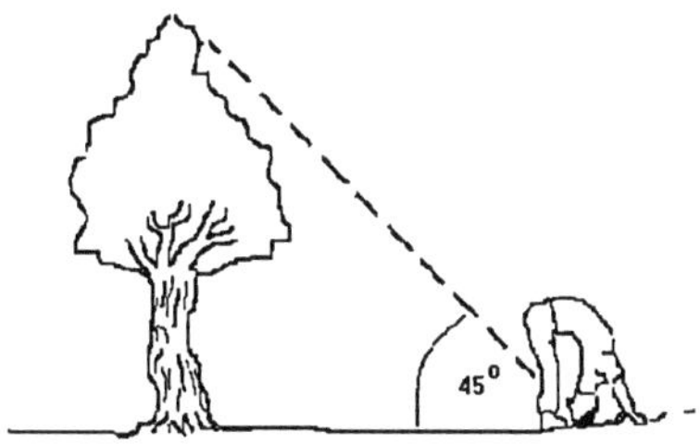

Figure 8 (from Surendran, 1996). A situated method of measuring the height of trees.

The use of epistemically interactive problem-solving routines illustrates another important principle of embodied cognition: the mechanism of *information self-structuring* through sensorimotor processes (Lungarella & Sporns, 2005; Pfeifer et al., 2007, Chapter 4.7; Clark, 2008, Chapter 1). The basic, but frequently ignored idea behind this principle is that organisms do not just passively sit and wait until they happen to be exposed to certain stimuli. Instead, they tend to actively induce meaningful patterns of sensory stimulation that dovetail with their neural information-processing routines. As a simple example, consider what happens when we move our fingers along the edges of a coffee cup. This exploratory activity is bound to generate systematic correlations between the sensory streams flowing through our visual, haptic, and proprioceptive channels. The "salient" information contained in these multimodal, time-locked correlations has been shown to enhance category learning and concept formation (Lungarella & Sporns, 2005). The cognitive "scaffolding" achieved through information self-structuring additionally relies on the natural morphology of our bodies, such as the clever positioning of sensors. For instance, the placement of touch sensors is especially dense on our fingertips, which is also the position that is naturally stimulated when we first grab an object with our hand. The morphology of our hands thus reduces the neural costs for controlling the acquisition of "good" data, which in turn simplifies the information-processing that has to be done by our brains. This cascade of cognitively potent interactions demonstrates how the Principle of Information Self-Structuring naturally complements the Principle of Ecological Balance.

Summing up, what does the suggested contrast between the *direct* vs. *indirect* cognitive significance of agent-environment interactions imply for the success of Rupert's gambit? Contrary to what Rupert seems to suggest, the main impetus behind embodied, situated, and/or extended approaches to cognition is not simply to label external representations (or any other resources) as "cognitive." Because of its focus on the information-theoretic implications of embodiment, the HEC gives us a principled methodological reason for why we should closely attend to the body-involving and world-involving loops of adaptive problem-solving. Rupert's HEMC-perspective, in contrast, invites the subtractive fallacy that we could always "in principle" study the hard ("cognitive") part of problem-solving by attending solely to the representational mechanisms inside the brain. In fact, Rupert (2009) repeatedly appeals to this move to show that our interactions with environmental resources can never play any direct cognitive role, because their contributions are always "screened off" by the organism's use of internal representations (see e.g., pp. 49, 55f, 120f, 127, 136, 139). It is thus hard to see how the research methodology mandated by HEMC differs in practice from the kind of methodological solipsism (Fodor, 1980) which is endemic to orthodox internalist conceptions of cognition.

This leads us to Premise (4) of Rupert's conservatism-or-simplicity argument, where it is argued that the proliferation of *cognitive systems* by HEC is gratuitous from an explanatory standpoint, and thus should be avoided. To deal with objection, let us turn to the use of dynamical systems theory in cognitive science, which has always been a main stimulant for the HEC.

3.3. Taking a "dynamicist stance" in cognitive science

The conceptual and mathematical apparatus of dyamical systems theory has long been the language of choice for control theorists. One of the earliest formal approaches to the topic, which spurred a lot of interest at the time, was James Clark Maxwell's (1868) analysis of the Watt governor – a centrifugal governor that can be used to control the speed of an engine by regulating the amount of fuel (e.g., steam) supplied. The point of treating something as a dynamical system is to be able to describe with great precision how the state of the system changes over time. This requires that we know the rules which govern the evolution of the system, such that every time we apply the rule to its current state, it will tell us the state of the system one step into the future. Being able to predict the future state of a dynamical system is clearly important if you're a control engineer. In many cases of interest, a dynamical system is mathematically represented as a set of variables that change continually, concurrently, and in interdependent ways over time in accordance with some rules that can be expressed as a set of equations (usually differential equations). More generally, the framework of dynamical systems theory provides a variety of concepts and tools for analyzing, and also visualizing complex patterns of change over time (Abraham & Shaw, 1992; Kelso, 1995).

It is easy to see why psychologists first became interested in dynamical systems theory to study problems of sensorimotor coordination (for a brief history, see van Gelder & Port, 1995; Chemero, 2009, Chapter 5). The coordination of sensorimotor behavior must be sensitive to the different temporal rates in which neural, bodily, and environmental processes take place. For instance, consider hitting a return of serve in tennis. Performing the necessary movements involves corresponding changes in the state of neurons which happen within milliseconds, processes of visual recognition which take several hundreds of milliseconds, the initiation and coordination of bodily movements which may continue for several seconds, and learning how to do it well takes many months and years of practice. In order to understand the complex dynamics of sensorimotor coordination, it is imperative that we have a framework for analyzing the continuous reciprocal interaction between heterogeneous components whose states change along multiple, often asynchronous time-scales. This is precisely where the apparatus of dynamical systems theory comes in handy.

As a classic example of a dynamical system, consider the case of two pendulums which, if mounted in close physical proximity on a wall, become swing-synchronized over time (Huygens, 1673; Salzman & Newsome, 1994). Dynamical systems theory allows an elegant formulation of this synchronization process (which is mechanically achieved via vibrations running along the wall) by treating the two pendulums as a single, *coupled* dynamical system whose temporal evolution can be given a succinct formal (but also graphical) representation. In general terms, the behavior of two entities (or subsystems) X and Y is said to be systemically coupled whenever there is continuous, reciprocal causal influence between the two, i.e. changes in the state of X continually influence, and are influenced by the concurrent state of Y. Technically speaking, the continuous, mutual interdependence between X and Y can be expressed in a pair of differential equations by treating the state variables of X in one equation as parameters for Y in the other equation, and vice versa. This makes dynamical systems theory an attractive toolkit for theories of embodied or extended cognition, because its methods are naturally suited for the analysis of complex causal interactions between components that lie on either side of the skin of a biological agent. By defining a set of coupling functions which systematically interweave the concurrent state changes of organismic and environmental components, dynamicists can provide compact, revealing descriptions of those interactions in terms of how the entire coupled system unfolds over time. According to the "dynamical stance" (Chemero, 2009, Chapter 4) in cognitive science, the ongoing, recicprocal, and often staggeringly complex causal interactions between neural, bodily, and environmental factors are best understood with the explanatory tools of dynamical systems theory, rather than with the familiar explanatory strategies of representationalist and/or computationalist approaches (for influential arguments in favor of the "dynamicist stance," see e.g., Varela, Thompson, and Rosch, 1991; Thelen & Smith, 1994; Kelso, 1995; van Gelder, 1995, 1998; Brooks, 1997; van Gelder & Port, 1995; Clark, 1997; Beer & Chiel, 1997; Haugeland, 1998; Keijzer, 2001; Thelen, Schöner, Scheier, & Smith, 2001; Beer, 2003; Thompson, 2007; Chemero, 2009).

Mathematically speaking, the behavior of nonlinearly coupled systems, i.e., systems whose dynamic patterns can only be described with the help of nonlinear differential equations, is *non-decomposable*. This means that unlike for linear systems, the evolution of their state space cannot, as a matter of principle, be modeled as collections of linearly separable components. However, dynamical systems theorists have developed alternative constructs to summarize the global patterns which are produced by the dynamics of nonlinearly coupled systems. A particularly useful practice is the definition of *collective variables*, whose values are determined by some relationship between individual system variables. For instance, in a classic study of the coordination of rhythmic finger movements by Haken, Kelso, and Bunz (1985), the phase relationship between the movements produced by a subject's two index fingers is described by a collective variable whose temporal evolution is shown to be determined by the frequency of oscillation (which acts as the *control parameter* of the dynamical system). With their detailed dynamic analysis of those patterns, known as the (original) Haken-Kelso-Bunz (HKB) model, they were able to show how complex patterns of interlimb coordination can spontaneously emerge from nonlinear interactions among the coupled component subsystems, without having to posit any centralized controller, or any task-specific neural "switching" mechanism that would trigger the observed state transitions.

A crucial feature of the dynamicists' practice of isolating control parameters, and modeling their influence on collective variables, is that the resulting mathematical characterization is surprisingly neutral with respect to the underlying mechanisms which are responsible for producing the observed patterns. As a case in point, the HKB-model of rhythmic motion not only applies to interlimb coordination, but also to a large range of mechanically diverse forms of behaviors all of which revolve around the coordination of nonlinearly coupled oscillators in living systems. For instance, it accounts nicely for the spontaneous phase transition that a horse undergoes when it switches, at a certain velocity, from a trot to a canter; the coordination of rhythmic movements with perceived features of the environment; various rhythmic aspects of speech recognition and speech production; the coordination of neural activity in different parts of the brain such as the synergetic effect of gamma oscillations in the 40 Hz band; the coordinated movements of socially coupled people; and even the temporal dynamics of cognitive problem-solving (for overviews with references, see Kelso, 1995, Chapters 2-3; Chemero, 2009, Chapter 5).

My goal for the remainder of this chapter is to take a closer look at the dynamicist stance in cognitive science, and to see if it warrants an argument for the HEC. To this end, my focus will rest on what I take to be the predictive and explanatory virtues of the dynamicist stance, not on what I consider to be some of its more controversial features. In particular, the dynamicist argument I shall defend does not require that we take on board all of the claims associated with the thesis of "radical embodied cognition" (Chemero, 2009), notably the claim that the use of the explanatory tools of dynamical systems theory is deeply incompatible with the explanatory role of representational and/or computational theories of cognition, and that cognitive science ought to completely discard the latter in favor of the former. My own ecumenical view, which is similar to the position defended by Clark (1997, 2001, 2008), is that dynamicist and representational (and/or computational) theories and models of cognition are better viewed as complementary rather than as mutually exclusive approaches to cognitive science. Another issue that I shall bracket for the purpose of our discussion is the current debate whether dynamicist models of cognitive and behavioral phenomena are intrinsically unsuited to provide genuine *explanations* of cognition. According to an influential line of criticism, the use of dynamicist models fails to explain cognitive processes, because those models do not tell us much about the mechanisms by which the dynamic properties of a complex system are causally produced by the interactions of its parts.[36]

3.3.1. A dynamicist model of minimally cognitive behavior

In this section, I want to showcase Randy Beer's (2003) sophisticated use of dynamical systems theory to analyze a situated, embodied model agent capable of "minimally cognitive behavior," i.e., the simplest behavior that raises questions of cognitive interest (ibid, p. 210). The minimally cognitive behavior examined in Beer's study is categorical perception, often considered as one of the "groundworks" of cognition (Harnad, 1987). In Beer's analysis, the nervous system, body, and environment of the agent are each conceptualized as dynamical (sub-)systems that are in constant interaction, with the goal of showing how the target behavior "arises from this interaction between these subsystems and can-

---

36 For influential criticisms of the dynamicist approach along those lines, see Dretske (1994), Eliasmith (1996), and Bechtel (1998); for a different view of the relationship between dynamicist and mechanistic views, see Zednik (2008).

not properly be attributed to any one component in isolation from the others" (ibid, p. 211). Following Beer's own presentation, I will first summarize the results of an earlier study of categorical perception in an evolved model agent (Beer, 1996), including some of Beer's psychophysical experiments. Then I will sketch the structure of Beer's multi-layered dynamical analysis of the observed behavioral patterns. The primary focus of my reconstruction is to illustrate the explanatory strategy that is implicit in Beer's dynamicist stance, rather than to detail any particular results.

There are several good reasons for picking Beer's model as our exemplar of choice. To begin with, Beer's focus on simple, idealized models of brain-body-environment systems allows him to undertake very detailed studies of the dynamics of such systems, and thus to capture the fundamental features of interest. We cannot hope to achieve the same level of detail with an analysis of real-world systems. First, it is beyond our experimental capabilities to manipulate, control, and measure not only large portions of an animal's nervous system, but at the same time all relevant states of its body and environment in real time. Second, even if we could somehow obtain those data, we would currently lack the analytic methods to apply the same methods of analysis to those data. In analogy to the history of physics, Beer (ibid.) compares his methodology of *computational neuroethology* (cf. Chiel & Beer, 1997) with the the study of "frictionless brains (and bodies and environments)" (ibid., p. 240), insofar as idealized models are better suited to uncover the most essential aspects of the phenomena under consideration. Another promising feature of Beer's model is that it extends the domain of dynamicist analyses beyond the realm of pure sensorimotor coordination. In a categorical perception task, an agent has to be able to sort continuous signals into a number of discrete categories (*labeling*), be able to distinguish better between members of distinct categories than between members within each category (*discrimination*), and respond differentially to objects from different categories (cf. Ehret, 1987). In Beer's model agent, these three subtasks are heavily intertwined, since the task of categorical perception is dynamically linked to moving in appropriate ways. As a result, there is no clean theoretical separation in Beer's analysis between the "cognitive" aspects of categorical perception and the "non-cognitive" aspects of sensorimotor coordination. Both aspects are modeled with the same underlying framework of dynamical systems theory.

In a previous study, Beer (1996) showed how evolvable model agents learned to discriminate categorically between circles and diamonds – catching the former, while avoiding the latter. Beer's embodied model agent is a (simulated) circular robot that can move back and forth along a horizontal line, while trying to discriminate between objects that are being dropped from above. The robot's self-generated sensory field consists of a bundle of seven "rays" by which it continuously perceives the vertical distance between itself and the falling object (Figure 9a). The robot is controlled by a 14-neuron, continuous-time, real-valued recurrent neural network (CTRNN) consisting of three layers: an input layer with seven sensory neurons, each of which receives signals from one "ray"; a layer of five interneurons, which are connected to the sensory neurons and also to one another, and which project to an output layer, consisting of two neurons, that drive the agent's two motors. On each learning trial, an object (either a circle or a diamond) is dropped from above the agent. Evolutionary algorithms were used to train the CTRNN so as to minimize the final horizontal separation for circles ("catching" response) and maximize the final horizontal separation for diamonds ("avoidance" response).

The most successful agent evolved to perform with an average success rate of over 99% on a set of evaluation trials. Beer found that the top-performing agent used an "active scanning" strategy to discriminate between circles and diamonds. Initially, the agent "foveated" a falling object by moving so as to bring it into the center of its field of view. It then actively "scanned" the falling object by repeatedly moving left and right. During the first half of the trial, the agent's movements are qualitatively similar for both categories of objects. They group into two bilaterally symmetric bundles, depending on which side of the agent the object initially appears. But at the second traversal of the two bundles, they begin to diverge: the agent either gradually decreases the amplitude of its scanning movements until it is centered (for circles), or it pauses briefly and then veers away (for diamonds).

Beer performed several "psychophysical" experiments on his model agent to confirm that it exhibited the defining features of categorical perception. In order to demonstrate labeling, the experimenters exposed the agent to a new class of hybrid objects whose shapes were smoothly interpolated between circles and diamonds. It was found that the agent's catch performance followed a characteristic S-shaped curve with relatively sharp transitions, thus reflecting the presence of a category boundary. To demonstrate discrimination, Beer investigated how well the agent would be able to catch a series of hybrid objects whose shapes were separated by some small, constant factor. The resulting performance curve is bell-shaped, with a peak discrimination capacity occurring as predicted near the category boundary, i.e., the region where adjacent objects are most dissimilar. Finally, probing into the agent's response to a variety of different shapes revealed that object width was the major determinant factor in discrimination.

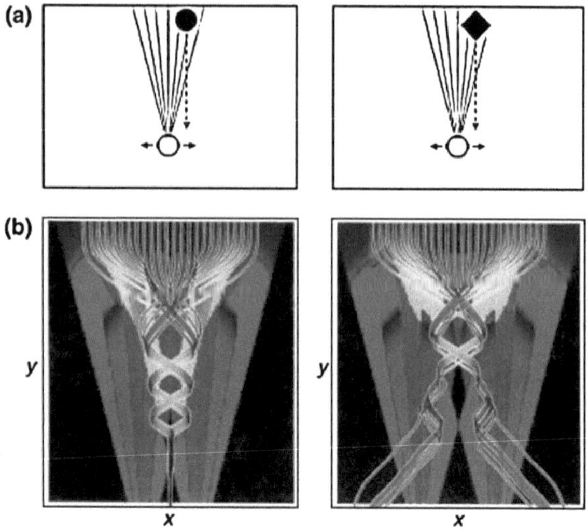

Figure 9 *(from Beer, 2000, p. 96). Dynamical analysis of an evolved model agent for catching circles and avoiding diamonds.* ***(a)*** *Basic setup of the categorical perception experiments. The agent can move horizontally while objects fall from above. It uses an array of seven "rays" to sense the distance to the object.* ***(b)*** *The movement of an evolved model agent for 24 circle trials (left) and 24 diamond trials (right). The horizontal position of the object relative to the agent is represented as x (with zero at the center), and the relative vertical position of the falling object (with zero at the bottom) is represented as y. The trajectories are color-coded according to the instantaneous horizontal velocity (blue = centering motion, red = avoidance motion), and superimposed on the steady-state horizontal velocity fields (see text below for explanation).*

Before we turn to Beer's dynamicist analysis, let me emphasize how Beer's robot agent displays some of the previously discussed features of embodied cognition. First, similar to the self-generated bodily twitches of Paul's robot (Chapter 3.2.3), the agent's active scanning movements are examples of *epistemic actions* performed to aid the process of categorical perception. Given the coarse spatial resolution of the seven "rays," it would be very hard for the robot to discriminate accurately between the two categories purely on the basis of static snapshots of its visual field, especially early on. Changing one's relative position to the object helps to enhance the quality of the visual cues that indicate differences in object shape. For instance, detecting the presence of a single corner on either side of the falling object (as opposed to a situation in which the same corner would directly face the agent) turned out to be enough to trigger avoidance behavior. Theories of embodied vision have similarly emphasized the constitutive importance of body orientation, posture control, head-turning, eye-movements, and foveation, for the purpose of visual processing (Churchland, Ramachandran, & Sejnowski, 1994; Ballard et al., 1997). The active exploitation of one's bodily structure and bodily actions for the service of perceptual processing is another demonstration of *information self-generation* (Chapter 3.2.4).

Second, there is no discrete instant in time at which the agent "makes its decision" to catch or avoid a falling object. A crucial feature of the agent's discrimination behavior, as we shall see, is that the robot repeatedly makes and unmakes its decision as its movements interact with the position of the falling object. For instance, switching the identity of the falling object (e.g., from a circle to a diamond) at different points during a trial reveals multiple sharp changes in performance. This shows that the agent wavers in its commitment to perform a certain kind of response, remaining open to reverse its behavior due to changes in the environment. In an important sense, the generation of a catch or avoidance response is a temporally extended process that emerges from a series of agent-environment interactions.

Let us now turn to the heart of Beer's paper, which is his detailed dynamicist analysis of the data we have described. The goal of Beer's analysis is to formulate a comprehensive model of the interactions between the movements of the robot's body, its changing relationship to the moving object, and the changing configuration of its neural circuits. The complete model consists of three "layers" of analysis. The purpose of the first layer is to characterize the dynamics of the entire coupled system that is formed by the evolved nervous system, the agent's body, and the falling object. Because the temporal evolution of the coupled brain-body-environment system is determined entirely by its own internal state (i.e., treated as a whole, it has no "inputs"), it forms a so-called autonomous ("closed") dynamical system. In contrast, the second and third layers of Beer's model are concerned with the relative contributions of neural, bodily, and environmental factors to the dynamics of the larger coupled system. Since the dynamics of nonlinearly coupled systems are not linearly decomposable, dynamicists have to apply special analytic techniques that effectively allow them to "cut through" the dense causal couplings, in order to isolate the effects of particular components they wish to study. By doing so, the individual component is thus treated as a non-autonomous ("open") dynamical system, and the goal of analysis becomes to show how its intrinsic dynamics interacts with the "input" received from other components to produce some specific pattern of the entire coupled brain-body-environment system (Beer, in press). In Beer's model, the second layer of analysis is concerned with showing how the dynamics of the coupled brain-body-environment system arises from the interactions between the nonautonomous agent (brain + body) and environment dynamics. Finally, the third layer of analysis is concerned with the nonautonomous dynamics of the evolved CTRNN (brain) as it is continually shaped by, but also shapes the range of sensory inputs and motor outputs. As it will become clear, each layer of analysis in Beer's model addresses different explanatory needs.

The first and most encompassing layer of Beer's analysis – the one to which externalists typically draw attention in support of the HEC – models the coupled system which spans the borders between CTRNN + agent + environment. The model describes a 16-dimensional dynamical system – i.e., a system whose evolution is uniquely described by a set of 16 interlocked differential equations. Of the 16 state variables which are represented, 14 designate the states of the 14 neurons, one variable (*x*) stands for the agent's horizontal distance from the falling object, and another variable (*y*) stands for the vertical distance of the agent from the falling object. Within this system, the agent's movements are determined by the states of the two motor neurons, which in turn are determined by the states of the five interneurons. The state of an interneuron depends on the states of the sensory neurons, as well as the states of the other interneurons, to which it is laterally connected. The states of the seven sensory neurons vary as a function of the agent's relative position with respect to the object (as expressed by the variables *x, y*) and the identity of the object. At the same time, any changes in the robot's motor neurons, which control the actuation of the robot's motors, change the robot's current location. This affects the value of the horizontal distance *x*, which changes the incoming sensory signals, which in turn change the states of the sensory neurons, and so forth. The dynamically coupled brain-body-environment system thus forms a feedback system, which makes it difficult if not arbitrary to disentangle "inputs" from "outputs," or causes from effects, because "effects 'play through the system' to become causes" (Beer, 2003, p. 236).

The dynamics of the entire 16-dimensional system is determined by the identity of the falling object, i.e., whether it is a circle or a cross. This means that the shape of the falling object is the only control parameter of the system. It determines the topology of the total state space, i.e., the set of all possible states in which the system can be in. Accordingly, Beer's mathematical model of the coupled brain-body-environment system as a 16-dimensional, 1-parameter, autonomous dynamical system describes all the possible ways in which the agent can move with respect to its environment (the falling object), and the relationship which each of those movements bears to the falling object and the agent's neural network. It can thus be used not only to describe the behavior observed in Beer's experiments, but to *predict* what an agent would do in all possible system configurations. Because of this feature, we can say that the dynamicist model expresses lawful, *counterfactual-supporting* generalizations about the agent's minimally cognitive behavior (Chemero, 2009).

Although it is not possible to visualize a 16-dimensional system directly, important insights into its coupled dynamics can be gained from studying carefully chosen lower-dimensional projections. It turns out that one particularly salient projection is formed by the combination of x, y, and the state of an interneuron (o_9). As their interactions unfold, the line of initial states is separated into two distinct bundles of trajectories that wind around one another several times, with peaks in o_9 occurring every time the agent scans to the right (or to the left, if we were to consider the activity of the symmetrical interneuron o_{11}). Plotting this interaction helps to detail the trajectories for circle and diamond trials, and how they are affected by gradually varying the shape of the object. For instance, in circle trials, the two circle trajectory bundles crossed over one another three times, then "collided," merged, and subsequently stayed in the midline; whereas for diamond trials, the peaks occurred earlier in time, with the bundles crossing only twice before diverging. In addition to describing the observed dynamics, Beer's model also helped to generate new, and subsequently confirmed predictions about the behavior of the system in hitherto unexplored scenarios, namely the occurrence of (additional) anomalous peaks in catching performance for objects that are wider than the original diamond (ibid, p. 226).

What accounts for the emergence of those features? In order to highlight the causes and effects of the agent's movements over time, the second layer of Beer's analysis decomposes the coupled agent-environment dynamics into (i) the effects that the relative positions of the object and the agent (x, y) have on the agent's motions; and (ii) the effect that the agent's motions have on the relative positions of the object and the agent (Figure 9b). Let me give a brief summary of some key results.

The only quantity which directly affects an agent's movements is the horizontal velocity ($\dot{x}$), which is determined by the states of the motor neurons. Since both agent and object are moving, we have to distinguish between the *instantaneous* horizontal velocity and its *steady-state* horizontal velocity. The steady-state velocity describes the velocity that an agent would adopt in the long run if we fixed a particular object at some location (x, y) in the agent's field of view. By fixing the location of an object, we specify a constant set of sensory inputs to the CTRNN, which means that if we waited long enough, the CTRNN would settle into one of its stable equilibrium points. In the diagram, these steady-state velocity fields are represented in the background. Regions of velocity which are directed towards the midline ("centering regions") are colored in blue, while regions of velocity away from the midline ("avoidance regions") are colored in red. Black regions of zero velocity are left black. Superimposed on the steady-state velocity field (using the same color-coding scheme) is the instantaneous horizontal velocity. Since agent and object are always moving, the sensory inputs are, of course, constantly changing rather than fixed. Since it takes some time for the neurons to respond to a change in inputs – a kind of "neural inertia" (ibid, p. 228) – this results in a lag between the internal state of an agent and the stable equilibrium points that are attracting it. Consequently, the instantaneous velocity lags behind the steady-state value.

As Beer's analysis shows, this lag turns out to be crucial for categorical perception. Consider the circle trials first (Figure 9b, left). As a result of neural inertia, the circle trajectory bundles repeatedly overshoot the midline, which produces the active scanning movements. However, they are always turned back toward the midline by the blue centering regions of the steady-state velocity field, until they eventually become trapped in the central black regions of no movement, so that catching occurs. In the diamond trials (Figure 9b, right), on the other hand, the central red regions of the steady-state velocity field are much stronger in the lower half of the plot, and also reach up higher. They push the diamond trajectory bundles away from the midline after the first scan, effectively keeping them from entering the black region in the center, which leads to avoidance behavior.

Finally, the purpose of Beer's third layer of analysis is to show how the described agent dynamics are produced by the evolved neural circuitry. For instance, how do the settings of the neuronal and synaptic weight parameters of the neural network generate the steady-state horizontal velocity fields? What properties of the network give rise to the central black region of no movement which has been shown to be critical for catching? To answer these questions, the neural architecture of the CTRNN is analyzed as a 14-dimensional, continuous-time, non-autonomous dynamical system that is continually affected by sensory signals. The goal of this analysis is to gain a deeper understanding of the internal neural dynamics which mediate between sensory inputs and motor outputs, in particular the subtle interplay between the steady-state velocity field and the the buildup of "neural inertia" that proved to be critical for discrimination.

Before we move beyond the details of Beer's case study to a general statement of the dynamicist argument for the HEC, let me point out two noteworthy features of Beer's dynamicist analysis. First, we have said that a specific state of the agent's environment – namely the identity of the falling object – is the only control parameter of the entire coupled system insofar as it determines the shape of the entire 16-dimensional brain-body-environment state space. Such a claim can easily be misconstrued as a fatal return to the "dark ages" of behaviorist stimulus-response models, which purported to explain intelligent behavior without making any reference to an organism's cognitively significant internal states.

However, a brief glance at the architecture of the CTRNN is enough to dispel this worry. Beer's categorizing robots are not purely reactive agents whose motor outputs depend *only* on the sensory input which they receive, but so-called dynamical agents whose behavior is in part determined by their *internal* states. This means that they are disposed to respond differently to the same environmental situation, in a way that is determined by their internal state, which in turn depends on the agent's past experience in that environment. Dynamical agents are also able to initiate and carry out non-reactive behaviors without the presence of environmental triggers and feedback, and can organize its behavior in anticipation of future events (cf. Beer, 2003, p. 238; see also Beer, in press).

Second, once the ties between dynamical systems theory and behaviorism have been severed, classical internalist cognitivists might argue that Beer's model of the evolved CTRNN (layer 3) is in itself a complete model of categorical perception. But this ignores the fact that an analysis of the CTRNN makes full sense only in combination with the two other, more comprehensive layers of analysis. Left to its own states, a previously inactive recurrent neural network would not be able to perform the object discrimination task. Contrary to standard feedforward networks, recurrent networks do not always generate the same output patterns for the same input patterns, because their current operation depends crucially on what other patterns it has previously received. The intrinsic context-dependence of a CTRNN on the history of its activity suggests that the neural network can only properly perform its task if it is attached to the particular body of a particular agent that is embedded in a particular environment. Without engaging in the right series of agent-environment interactions, there would be no task-relevant background activity to turn on the CTRNN. As a partial explanation of the minimally cognitive behavior of the embodied agent, the neural network dynamics revealed by Beer's model of the CTRNN (layer 3) is thus useful only because we already have a model of the coupled dynamics of the brain-body-environment system (layer 1), and how it is generated by the interaction between agent dynamics and environment dynamics (layer 2). Hence it would be inaccurate to say that the discrimination task is solved by the nervous system alone, as we shall now argue.

3.3.2. A dynamicist argument for the HEC

Let us now move beyond Beer's analysis of categorical perception to a more general discussion of the implications of dynamicist models in cognitive science. Contingent on their empirical success, the use of of dynamical systems theory to explain various aspects of intelligent behavior seems to lend support to the following (still enthymematic) argument for the HEC:

1. Understanding the dense, continuous, and often nonlinear causal interactions in which neural, bodily, and environmental factors conspire to generate intelligent behavior requires dynamicist explanations.
2. The success of dynamicist explanations requires that we treat the causal interactions among these factors as part of a single, coupled dynamical system.

C1: *Therefore*, understanding the complex ways in which neural, bodily, and environmental factors conspire to generate intelligent behavior requires that we treat them as part of a single, coupled dynamical system.

C2: *Therefore*, understanding the complex ways in which neural, bodily, and environmental factors conspire to generate intelligent behavior requires that we treat them as part of an extended cognitive system.

Several comments are in order before we'll get to the crux of the argument, which is meant to establish the existence of extended cognitive *systems* (C2). If the HEC is formulated primarily in terms of extended cognitive *states* and *processes*, an additional inference would be required to derive the HEC from the existence of extended cognitive systems (for a criticism of this extra inference, cf. Adams and Aizawa, 2008, Chapter 7; see my Chapter 5.5. below for a response). However, I have decided to focus on the notion of an extended cognitive *system* here, for a couple of reasons. For one, the dynamicist stance in cognitive science traffics primarily in systems. Hence it is an interesting project all by itself to understand the ontological commitments of systems-centered dynamicist explanations. Second, I shall argue that an analysis of intelligent behavior in terms of extended cognitive systems is indeed sufficient to vindicate at least some version of the HEC. My view is indirectly supported by the fact that Rupert (2009) devotes an entire chapter to refute dynamical systems-based arguments for the HEC. One of the reasons why Rupert is committed to contest the dynamicist argument is that he is independently attracted to a "systemic" mark of the cognitive. According to Rupert (2009, Chapter 3.2), a state or process is cognitive just in case it is part of a cognitive system (see my Chapter 6.2 below for further discussion). Consequently, the target of Rupert's critique is clearly C2, rather than the separate inference from cognitive systems to cognitive processes that has been critizised by Adams and Aizawa. In what follows, I thus plan to focus on the argument for C2.

I also take it that the argument for the first conclusion (C1) is perfectly valid as it stands. As for its soundness, I don't pretend that my brief sketch of the dynamicist stance has provided anything like conclusive evidence for Premises (1) and (2). As I said earlier, the main goal of this chapter is not to defend the dynamicist stance as such, but rather to explore its implications for the HEC. In light of these considerations, I ask the reader to accept both premises at least provisionally.

Critics of the HEC are therefore mostly likely to object to the suggested inference from C1 to C2. For even if we grant that dynamical systems-based explanations are valuable and in many cases practically indispensable tools for cognitive-scientific research, it is by no means obvious that we should consider a dynamically coupled brain-body-environment system as an extended cognitive system. There are two different reasons for being skeptical of such an inference, depending on which point is taken to be the relevant bone of contention. On the one hand, one could ask: (1) why should we consider a dynamically coupled brain-body-environment system as a single cognitive *system*? On the other hand, we could also ask: (2) why should we consider a dynamically coupled brain-body-environment system as a single *cognitive* system? (Or perhaps, if we don't believe in rigid distinctions, what determines the gradient towards *cognizance* in dynamical brain-body-environment systems?) Both questions have come up in philosophical discussions of the dynamicist stance, so it is only natural to review some answers which have been suggested.

Let us begin with question (1). The conceptual analysis of what features make a dynamical system *cognitive* is often characteristically side-stepped by many proponents of the dynamicist stance.[37] To an extent, this eschewal is motivated by their deep skepticism about the usefulness of traditional constructs such as intentionality, mental representations, and also consciousness, which have often been cited as putative "marks of the cognitive" (see Chapter 2). As opposed to the representation-laden mentalistic framework of the "intentional stance" (Dennett, 1971) – roughly speaking, the framework of belief-desire psychology – proponents of the dynamicist stance place a distinctive emphasis on concrete aspects of embodied and situated behavior. Accordingly, their implicit "mark of the cognitive" often boils down to little more than the coarse-grained functional role of cognition as a mediating factor between perception and action.[38] For instance, Beer (2000, p. 97) describes the working hypothesis of the dynamicist stance as the claim that "through increasingly sophisticated uses of internal state to mediate between perception and action, more cognitive behavior emerges from the dynamics of situated action." Reflecting on what it takes to confirm

37 For interesting dynamicist attempts to define *cognition* in terms of general organizational properties of living systems, see e.g., Di Paolo (2005), and Barandiaran & Moreno (2006).

38 Compare, for example, the relevant discussions by Thelen & Smith (1994), Brooks (1999), Beer (2003), and Spivey (2007).

their hypothesis that cognitive systems are a subset of dynamical systems, van Gelder and Port (1995) point out that "If we are interested in *cognitive* systems, then the behaviors of interest are their *cognitive performances* (perceiving, remembering, conversing, etc.), and it is *these* behaviors, at their characteristic time-scales, that must unfold in a way described by the rule of evolution" (ibid, p. 11). Van Gelder (1998) makes a similar point when he notes that "[t]his paper simply takes an intuitive grasp of the issue for granted. Crudely put, the question here is not what makes something cognitive, but how cognitive agents work" (ibid, p. 619). The same confidence in the working psychologist's largely pre-theoretical grasp of cognitive phenomena is evident in Beer's (1996, 2003) paraphrase of "minimally cognitive behavior" as "the simplest behavior that raises cognitively interesting questions."

Since I have already expressed my reservations about the presupposition that there must be one or more essential "marks of the mental" (Chapter 1.4), I applaud the dynamicists' pragmatism if it leads to good psychological models. Consequently, I shall not try to improve upon the suggested understanding prior to, and independently of the discussion of any concrete psychological models. In the previous section, we already encountered one such model – namely Randy Beer's (2003) model of categorical perception as a model of minimally cognitive behavior. As a way of bypassing philosophical worries about any generic "cognitive mark" of dynamical systems, let me therefore spell out more succinctly in what sense we may consider Beer's brain-body-environment system as an extended (minimally) cognitive system.

From the outset, we should stress that it makes little sense to adjudicate the existence of dynamical systems *simplicter*, in abstraction from any specific phenomena ("patterns") of interest whose occurrence we seek to explain. For instance, astronomists refer to the solar system because they want to explain certain patterns of planetary motion around the sun. Similarly, what motivates psychologists to use the tools of dynamical systems theory is the observation of certain patterns of intelligent behavior. Following the above suggestion by van Gelder and Port (1995), we can say more generally that a dynamical system S is of a type Φ just in case the temporal evolution of S is responsible for generating behavioral patterns of Φ. In Beer's model, S is taken to be a coupled brain-body-environment system, and Φ is taken to be a certain type of (minimally) cognitive behavior, namely the performance of categorical perception.

What specific aspect of (minimally) cognitive behavior does Beer refer to with his notion of "categorical perception"? It is tempting to say that he is referring to the behavior *of* a non-extended agent here, namely its ability to catch circles and avoid diamonds. At least sometimes, Beer seems to have this interpretation in mind. For instance, he defines an *agent* as the coupled brain-body subsystem and the *behavior* of an agent as its trajectory of motor actions (Beer, in press, p. 16). However, this is quite evidently not the same notion of *behavior* that Beer seems to have in mind when he claims that "behavior is a property of the entire coupled brain-body-environment system, and cannot in general be properly attributed to any of the subsystems in isolation from the others" (ibid., p. 16f; see also Beer, 2003, pp. 211, 235f). Recognizing this discrepancy, Beer himself (in press, p. 16f) rightly points out that there are two distinct explananda here. On the one hand, there is the agent-side dynamics, which includes the trajectories of the agent's motor actions. But on the other hand, those trajectories are merely projections of the complete trajectories of the entire brain-body-environment system, which are the ultimate explanatory targets of his dynamicist framework. Properties which can truly be attributed only to the agent-environment system as a whole must show up as resulting from those complete trajectories. In Beer's model, an example of such a systems-level property would be *the horizontal separation between the agent and the object at the end of a trial* (Beer, 2003, p. 236). Importantly, this is a relational property of brain-body-environment systems, because its characterization essentially refers to a certain (changing) *relationship* between the agent and its environment, rather than to any intrinsic behavioral property of an agent (e.g., the activation of the robot's motors). To see that the distinction between intrinsic and relational properties is not merely a logical nicety in this context, consider the Darwinian evolutionary algorithm which Beer used to train the CTRNN of his model agent. The global performance measure which the algorithm had been set to optimize was to minimize the final horizontal separation for circles, and to maximize the final horizontal separation for diamonds. This shows that evolution indeed selected for a behavioral outcome that implicitly involves a particular *relationship* between agent and environment – a relationship which, as Beer has shown, emerges from the dynamics of a coupled brain-body-environment system.[39]

Let us now turn to the second question mark behind the suggested inference from C1 to C2 – i.e., the one which concerns the nature of cognitive *systems*. Is there anything like a "mark of the systemic," in particular when it comes to the

---

39 An alternative interpretation, to be discussed in much greater detail in Chapter 6.3, is to consider the capacity for categorical perception as a property of an agent which has a wide realization (Wilson, 2001; 2004). For instance, in order to determine the biological fitness of an organism, we have to look beyond its spatiotemporal boundaries and determine how it functions causally in its ecological niche. Metaphysically speaking, the possession conditions for a widely realized property do not logically supervene on the intrinsic properties of its bearer, because the way in which the property is individuated directly implies that the individual bears the right kind of relationship to its environment. As I shall argue in Chapter 6.3, this interpretation equally implies the HEC.

application of dynamical systems theory to the domain of psychology? In their previously cited essay on the "dynamicist hypothesis" in cognitive science, van Gelder and Port (1995) also brush up against the question of how to draw the boundaries of a dynamical system. As the key criterion for the individuation of a *system*, they point to the right causal organization among a set of elements: "for a set of aspects to qualify as a system, they must be interactive and self-contained: change in any aspect must depend on, and [for relevant purposes] only on, other aspects in the set" (ibid., p. 5). In their view, this causal criterion correctly includes – at least to a good degree of approximation – the orbit of the planets in our solar *system*, but it excludes, again to a good degree of approximation, the haphazard selection consisting of the position of my pencil and the color of my car (ibid.).

In the remainder of this section, I want to reflect a bit further on the suggested notion of a (cognitive) *system*, mainly against the backdrop of Rupert's portrayal of the theoretical alternative between HEC and HEMC (cf. Chapter 3.3). Recall that the goal of Rupert's dilemma-style argument against the HEC ("Rupert's gambit") was to show that we ought to prefer the HEC over the HEMC on grounds of either theoretical *conservatism* (Premise 3) or theoretical *simplicity* (Premise 4). Since I have already defended the HEC against the charge of conservatism (Chapter 3.3), I now want to scrutinize Rupert's reasoning behind the simplicity charge which is couched as Premise (4). According to Rupert, the HEC violates the principle of "Occam's Razor" insofar as

4. If the HEC recreates the "standard" internal cognitive architecture in its models, and *also* posits additional extended cognitive systems, then we should prefer HEMC over HEC on grounds of simplicity.

In support of Premise (4), Rupert (2009) claims that HEC-style explanations involve all of the entities and categorical distinctions that are invoked by the internalist HEMC, but in addition the HEC gratuitously takes on an extra ontological commitment "by joining elements from the various categories and dubbing the result a 'cognitive system,' even though the introduction of this entity serves no cognitive-scientific purpose" (p. 20; see also 2010, p. 347). Thus HEC seemingly multiplies the number of cognitive systems beyond necessity.

My response to this objection is that once again, Rupert's characterization of the debate reveals a subtle yet important misconstrual of the HEC. In particular, it makes the ontological commitments of the HEC appear more puzzling than they really are. As Rupert is well aware (2009, p. 18), simplicity decisively affects our theory choice only if two theories yield structurally equivalent explanations. But there is at least one important aspect in which HEC and HEMC diverge, which concerns their preferred ways of slicing up the relative functional contributions of our brain-bound neural machinery vis-à-vis the bodily and environmental context in which it is embedded.

To gain a decisive edge by considerations of parsimony, Rupert has to show two things: first, that the structure of HEC-explanations can always be neatly decomposed into an organism-bound portion that is cognitive, and an external portion which "merely" reflects the causal influence of the environment; second, that the structure of this internal portion simply "apes" (ibid., p. 41) the corresponding HEMC-explanation of organism-bound cognition. The second requirement is particularly important to sustain the charge of redundancy. For Rupert, the common element which he believes to be shared by HEC and HEMC is what he calls the "standard" cognitive architecture – an invariant (albeit maturing) bundle of self-contained, brain-bound capacities whose internal cognitive functioning remains largely unaffected by being embodied and embedded in various environments. Consequently, the HEC-theorist can avoid Rupert's charge of redundancy by showing how the cognitive functioning of our brain-bound machinery can be deeply transformed through its intimate causal commerce with the body and external "wideware." The resulting difference is a matter of degree and explanatory emphasis, but no less real for that matter.

By way of analogy, consider first the familiar eukaryotic cell, which is the fundamental building block of animals and plants. According to the endosymbiotic theory of the origin of mitochondria and chloroplasts (Margulis, 1970), both organelles were originally bacteria that became engulfed into the cytoplasm of prokaryotic cells. Once incoporated, they entered into symbiotic relationships with the surrounding cell, as a result of which they evolved in their function as power generators for the prokaryotic cell, performed either by means of cellular respiration (mitochondriae) or by photosynthesis (chloroplasts). Due to their specialization, the incorporated entities became functionally dependent on the metabolic regime of their cellular environment, with the effect that they eventually lost their ability to survive as independent units.

Why do we consider the eukaryotic cell as a biological system of its own that contains the organelles among its proper parts? The answer, to be spelled out further below, is that its elements are causally interdependent in ways that reliably support new types of systemic functions and operations that are qualitatively different from those performed by individual elements in the system. Examples of how structured collections of individuals evolve into new kinds of individuals are by no means restricted to the evolution of eukaryotic cells from prokaryotic cells, but have occurred in the evolution of multicellular organisms from unicellular organisms, or the evolution from solitary to social species. A similar process of functional incorporation motivates the idea of extended cognitive systems. More precisely, we shall say that the HEC is ontologically committed to extended cognitive systems in explanatory contexts where organismic and environmental resources and/or processes are *coupled* (in their normal operation) so as to form *integrated cognitive systems* with *functional gains*. Following a related proposal by Wilson (2010b, p. 285), we can unpack this complex characterization as follows (see also Clark, 2008, p. 33). First, two (or more) elements are coupled just in case they reliably enter two-way causal connections. Second, coupled elements form an integrated system in situations in which they function as a single, coherent entity within the causal nexus – with causes affecting the resultant system as a whole, and the activities of that system as a whole producing certain effects. Generally speaking, this implies that the density of connections and dependencies among the elements of a system significantly outweighs the density of connections and dependencies between elements inside and outside of the system. Third, an integrated system is cognitive just in case it manifests certain cognitive properties (states, processes, capacities). Fourth, an integratively coupled cognitive system shows functional gain just when it either enhances the existing cognitive functions of its coupled *elements*, or manifests novel cognitive functions as a *whole* relative to those possessed by any of its elements.

Distinguishing between the functional gains of wholes and their parts is important in the context of Premise (4), because it shows that evidence for the existence of extended cognitive systems can be found at multiple levels of explanation. At the system-level, the interesting cognitive properties of integratively coupled systems are often highly organization-dependent or *emergent* in the sense of Wimsatt (1986). Wimsatt's conception of emergence is defined against the backdrop of his notion of "aggregativity." A property P of a complex system S is said to be *aggregative* if and only if (i) P(S) is invariant under the intersubstitution of parts of S; (ii) P(S) remains qualitatively similar (differing only in value) under the addition or subtraction of parts; (iii) P(S) is invariant under the decomposition and re-aggregation of parts; and (iv) there are no cooperative or inhibitory interactions among parts. Correspondingly, a property is said to be *emergent* to the extent that it fails one or more of the suggested conditions of aggregativity. This makes emergence a matter of degree, rather than an all-or-nothing phenomenon. Generally speaking, we can say that emergent cognitive properties resist the familiar explanatory strategies of decomposition and localization (Bechtel & Richardson, 1993; Clark, 1997, Chap. 6). This means that we cannot hope to explain the production of emergent cognitive outcomes by first dividing up the entire system into a number of discrete parts, characterizing the contributions of those parts as if they were isolated from each other, and then simply adding up their outputs. Consequently, HEMC-style decompositions of extended cognitive systems are bound to be most revealing for aggregative cognitive properties, and least attractive for maximally emergent cognitive properties. The wide range of cases which lies on a continuum in between can only be judged by the explanatory virtues of specific cognitive-scientific models, not by controversial intuitions about simplicity.

Extended cognitive systems also leave a stamp on their parts. We do not strictly have to require that the pre-existing cognitive functions of integratively coupled components must always be enhanced. Since internal and external resources may trade off performance gains as a result of becoming elements of an extended cognitive system, the pre-existing cognitive functions of internal elements may at times be impaired. What matters is that the components of an integrated cognitive system typically behave differently as elements of the system than they do in isolation, and in ways that often cannot be predicted from the laws governing the behavior of those parts in simpler configurations (cf. Boogerd et al., 2005).

Beer's (2003) detailed analysis of how the network dynamics of the evolved neural controller is continuously affected by, and also continually affects the changing interactions between agent and environment is a striking demonstration of this transformation. As several other, in-depth analyses of idealized brain-body-environment systems have similarly shown, the nonautonomous neural dynamics of an "embodied" brain are different from the intrinsic dynamics of an "envatted" brain considered in isolation; and similarly, the nonautonomous dynamics of an environmentally "embedded" agent (i.e., a coupled brain-body subsystem) differs from the intrinsic dynamics of a "disengaged" agent considered in isolation of its environment (Husbands, Harvey, & Cliff, 1995; Chiel, Beer, & Gallagher, 1999; Beer, Chiel, & Gallagher, 1999; Beer, 2003). Let me round out my discussion with a few intuitive examples of the causal effects which the dynamic organization of a complex system has on the behavior of its parts.[40]

For instance, as Sten Grillner's long-term research on lamprey eels has shown, the central pattern generators in the spinal cord of a lamprey operate at different frequencies when they are part of an isolated spinal cord compared to the situation where they are normally integrated into the animal's body (Grillner, McClellan, & Perret, 1981). A detailed biomechanical model of a swimming lamprey failed to generate appropriate contractions along the animal's body when the hydrodynamics of water were factored out of the equation. The inappropriate patterns of movement predicted by the model were later observed when real lampreys were removed from the water and induced to swim (Williams et al., 1995). Similarly, a realistic model of a leech body failed to produce

40 The following examples are cited after Chiel and Beer (1997).

the effective crawling movements observed in intact leeches when the same leech body was stimulated with motor signals generated by artificially isolated nervous systems (Kristan et al., 1996). This shows that without the constant sensory feedback provided by the animal's own movements, the nervous system was unable to generate motor patterns that support adaptive behavior.

A considerable leap upwards on the evolutionary ladder, the cognitive scaffolding of our biological brains by our artifact-enriched environments provides another tentative example of the suggested transformation of an agent's internal cognitive resources by its natural and/or cultural environment (Chapter 1.3). The enourmous cortical plasticity of our "spongy" brains makes us uniquely apt to appropriate linguistic, representational, and technological artifacts into our biologically pre-disposed cognitive capacities, and thereby enhance the cognitive functions that our biological brains support (Donald, 1991; Dennett, 2000; Clark, 1998, 2003, 2006a, 2008; Menary, 2007; Sutton, 2010). In support of extended cognition, HEC-theorists have argued that the skillful incorporation of cognitive artifacts into our problem-solving routines is accompanied by corresponding changes in our neural architecture that reflect our increasingly tacit, automatic reliance on their bio-external availability (cf. Chapter 4-5 below for further discussion). It remains an area of active research to determine how deeply our artifact-rich environments sculpt, enhance, and transform our internal cognitive profile (Baltes, Reuter-Lorenz, & Rösler, 2006; Mareschal et al., 2007a; 2007b). My goal here is not to settle this empirical question, but merely to provide a rationale for why the outcome of this kind of research should have a significant bearing on how we individuate cognitive systems. This is because the developmental scaffolding of our bio-*internal* cognitive dynamics constitutes evidence against Rupert's hidden assumption of an invariant "standard cognitive architecture" that is essentially independent of the ways in which our brains are embodied and culturally embedded.

3.4. The importance of working downwards

In his recent book, Rupert (2009, Chapter 7) has sharpened his criticism of the dynamicist argument for the HEC. He suggests that dynamical systems-based models of brain-body-environment systems can be understood conservatively as mathematically elegant expressions of the fact that organisms are sensitive to changing values in their inputs, and modulate their outputs accordingly so as to complete certain cognitive tasks. Such an understanding is compatible with the HEMC. It does not imply that we ought to consider systems with the right dynamically-coupled causal structure as extended cognitive systems in their own right. In a compressed statement of his criticism, Rupert recapitulates his "deflationary" (ibid., p. 137) reading of the dynamicist stance as follows:

> "The human organism has psychological or cognitive capacities to interact with its environment in various ways, by coupling with it; [...] and in doing so, the organismically bounded cognitive system carries out its cognitive work by interacting with external materials. [...] Cognitive psychology hopes to explain the abilities of cognitive systems to interact with their environment. Cognitive psychologists would like to know how the organism acquires such abilities and how these abilities are manifest in different ways in different circumstances, and a dynamical-systems-based approach may provide the best means for doing so. This, however, is a science of non-extended systems interacting in complex ways with their environments" (Rupert, 2009, p. 137f).

Rupert's deflationism is based in part on his confidence in the method of functional analysis which has been germane to "orthodox" research programs in cognitive psychology (cf. Rupert, 2009, p. 37). According to this method, cognitive psychologists proceed in roughly the following manner: first, they identify the presence of a certain cognitive capacity by its functional role (e.g., categorization, planning, perception, memory, language-use, or inference); second, they break down the capacity into a number of subfunctions; third, they try to discover the supporting mechanisms whose operation causally explains token manifestations of those capacities on a given occasion.[41] The method of functional analysis is an instance of the more general pattern of "decomposition and localization" strategies for explaining the behavior of complex systems (Bechtel & Richardson, 1993). An implicit commitment to this strategy galvanizes Rupert's aforementioned claim that the primary goal of cognitive psychology is to discover the underlying, and contextually invariant features of the brain-bound cognitive system which allow the organism to couple with its environment.

However, this is hardly the explanatory project in which psychologists working in the dynamicist tradition take themselves to be engaging. Consider, once again, Beer's (2003) three-layered dynamicist model of categorical perception. Beer did not try to understand the dynamics of the coupled brain-body-environment system by *first* describing the inherent properties (and dynamics) of isolated components, and *then* modeling the global behavior of the complex system as an additive re-aggregation of those individual contributions. Instead, his method of analysis follows a path which is importantly different from traditional decomposition-and-localization strategies. He *first* aims to discover lawful generalizations about the dynamics of coupled brain-body-environment systems, and *then* applies various techniques to analyze how individual components conspire in real time, and in specific contexts to produce the global patterns of interest.

---

41 Classical statements of the method of functional analysis can be found in Dennett (1978), Marr (1982), and Cummins (1983).

As I have argued (Chapter 3.3.2), the dynamicist's departure from traditional decomposition-and-localization strategies has two related reasons. First, many of the interesting integratively coupled systems are nonlinear dynamical systems, and thus at most minimally decomposable. Hence they are intrinsically unsuited for the method of functional analysis advocated by Rupert. Second, the individual components of an integratively coupled dynamical system behave differently as elements of the complex system than they would if considered in isolation. For instance, the neural activity of the CTRNN (in Beer's model agent) is critically time-dependent on a particular history of agent-environment interactions. Hence there is no single, discrete cognitive function of the CTRNN that is invariant across the temporal evolution of its state space.

These considerations indicate a stark epistemological contrast between decompositional and dynamicist "styles" of explanation (cf. Clark, 1997, Chapter 6). This contrast is directly related, of course, to the ontological property of complex systems behavior that we have called (following Wimsatt, 1986) *emergence*. Since emergence (in this sense) is a matter of degree, we can say that the expected rate of success for decompositional explanations of some psychological phenomenon *P* is directly proportional to the degree of "aggregativity" of *P*. This is because for highly aggregative systems, we we can approximately decide to ignore the mutual effects of the interactions among individual parts in describing their individual contribution to the system as a whole; for highly emergent systems, the effects of those interactions cannot possible be ignored. Because of the suggested contrast between dynamicist and decompositional explanations, I shall now argue, we ought to be wary of Rupert's gambit to siphon off the instrumental benefits of the dynamicist stance without taking seriously its ontological commitments.

For expository purposes, it will be helpful to render the contrast between prototypes of "pure" decompositional and "pure" dynamicist explanations as vividly as possible. To that end, let us conceive of research strategies in terms of business cycles. (After all, it is a widely considered mark of the scientific that science is in the "business" of delivering explanations, even though there is there is no consensus about what constitutes a good explanation.) Each business cycle involves three steps. The goal of Step 1, the investment stage, is to bring a suite of tools to the explanatory task at hand. Stage 1 can thus be seen as an epistemic investment that is put to work in stage 2, and will hopefully pay epistemic dividends by the time we get to Stage 3. Employing this tripartite scheme, we can portray the difference between decompositional and dynamicist styles of explanation as a general contrast between what I shall call "bottom-up" and "top-down" approaches to problem-solving.

A. Decompositional "bottom-up" approaches

1. [*invest*] Decompose a complex system into its constituent parts.
2. [*labor*] Describe the intrinsic properties of these parts when considered in isolation, the relationships among those parts, and then the interactions which they undergo as a result.
3. [*payoff*] Assemble the parts accordingly to account for the behavior of the complex system.

Within cognitive science, the classic research strategy known as *homuncular functionalism* (Dennett, 1978; Cummins, 1983; Lycan, 1987) conforms to a decompositional scheme of explanation. Homuncular functionalism tries to explain the intelligent behavior of a complex system (e.g., a person, an animal, a computer program) by dividing it up into a hierarchically organized series of relatively modular components ("homunculi"), each of which exhibits an increasingly simpler (and less "intelligent") functional organization than the system as a whole. The goal is then to iterate this process of functional decomposition long enough, all the way down to the level of relatively simple components whose behavior can be directly implemented by basic physical or mechanical resources (e.g., logic gates, connectionist processing units, neural ensembles). Finally, once that level has been reached, we have effectively "discharged" the homunculus, because now we can explain "bottom-up" how the intelligent behavior of the complex system arises from the composition of its parts. This kind of research methodology makes homuncular functionalism a *synthetic* version of decompositional explanation (Bechtel & Richardson, 1993, Chapter 2): first, we conjecture how the behavior of the complex system might be performed by a set of subfunctions; second, we localize the components which are responsible for performing those subfunctions.[42]

B. Dynamicist "top-down" approaches

1. [*invest*] Assemble a collection of interacting parts into a complex system.
2. [*labor*] Describe the dynamic patterns of the emergent complex system in (considerable) abstraction from the intrinsic properties of its parts, and also from the mechanism by which those patterns are produced.
3. [*payoff*] Slice up the complex system into a number of elements, and use the dynamic analysis of the whole to account for the distinctive behavior of those elements as interacting parts of the complex system.

---

42 We use an *analytic* strategy of decomposition when we first physically isolate individual components within a system, and then determine what each component does (as, e.g., in classical chemistry).

Many of the "pure" dynamicist explanations can be subsumed under this "top-down" approach to problem-solving. It is particularly well-suited to understand the dynamics of integratively coupled complex systems, in which the behavior of individual elements is greatly affected by their participation in the system as a whole. As we have shown for layers two and three of Beer's model, an important goal of dynamicist modeling is to analyze the relative contributions of different subsystem dynamics *through* a proper dynamic understanding of the coupled system dynamics. As suggested by the title of this section, I refer to this explanatory strategy as the *method of working downwards*.

Equipped with a basic understanding of the epistemological contrast between decompositional and dynamicist styles of explanation, I now want to spell out my critique of Rupert's deflationism in greater detail. My discussion will be focused on Geoffrey Bingham's (1988) interesting dynamicist paradigm for studying the performance of complex human actions.

3.4.1. Brains, bodies, and task-specific devices

Geoffrey Bingham (1988) introduced the notion of a *task-specific device* (TSD) into human movement science, and proposed it as a centerpiece of his research into the dynamics of complex human actions. His integrative research methodology combines "bottom-up" investigations (e.g., through biomechanics, physiology) of the intrinsic resource dynamics of individual body parts with "top-down" investigations of the dynamically constrained (e.g., through motor control) organization of human actions.

As an illustration of a typical research topic in movement science, consider the action of throwing a football as far as possible (Bingham, 1988, pp. 242f). Performing such a throw requires that the kinetic energy of the football is maximized at the moment of release. As a way of overcoming the natural limitations of our musculo-skeletal system, skilled throwers employ a combination of two behavioral strategies. One strategy is to accumulate kinetic energy in the more massive trunk and limb segments, and subsequently pass it on to the object through the smaller limb segments in the throwing arm. The other strategy is to store potential energy in the tendons and muscles of the wrist through prestretching, and then recover that energy just prior to the release. More generally speaking, finding the optimal throwing motion for a given purpose is determined by several task-specific dimensions. For instance, people use different types of throwing motions depending on the mass and size of the projectile. For relatively light projectiles (e.g., a baseball), we tend to use an overhand throw which relies heavily on wrist and finger flexion; for projectiles around 500 grams (e.g., football), we tend to employ a lob-style of overhand throwing, in which we rotate our shoulder and keep wrist and elbow relatively fixed; for heavier projectiles (e.g., metal balls), we typically use a distinctive shot-put motion. For the selection of a particuclar throwing motion, the trade-off between power and accuracy must also be taken into account (ibid.)

The *human action system* (HAS) consists of four main subsystems – the link-segment system, the musculo-tendon system, the circulatory system, and the nervous system. Each style of throwing involves a different combination of bodily resources which the HAS has to bring into distinctive dynamical configurations. To perform a skilled throw, the HAS must take into account the available dynamic resources relative to the dimensions that ought to be optimized, so that it can discover, assemble, control, and maintain their appropriate organization in real-time. Understanding this difficult task poses two related theoretical challenges for models of the HAS (also known as the "degrees of freedom" problem; cf. Fowler & Turvey, 1978; cited after Bingham, 1988): first, to provide a principled description of the degrees of freedom that are associated with the dynamic properties of its various material components; second, to explain how the large number of degrees of freedom are constrained in the performance of an action so that its dynamics can be captured (and thus effectively controlled) in terms of relatively few degrees of freedom.

The degrees of freedom problem presents a formidable challenge for movement science. First, because of the sheer number of structural components involved, enumerating the degrees of freedom for the HAS includes a potentially very large number of dynamic properties that vary in form, scale, and type. Second, performing such a dynamic analysis is difficult because of the rampant nonlinearity of the interactions between the different component dynamics that form part of the HAS.[43] As a result, it is impossible to explain the dynamic organization of the HAS simply by adding or superimposing the inherent dynamics of its individual components. Conversely, accounting for the dynamics of individual components from the global action dynamics presents a mathematically difficult inverse problem which cannot be solved by simple subtractive procedures (ibid., pp. 226f). Third, in order to achieve and maintain a stable, and functionally effective organization of action, the HAS needs to factor in the additional "incidental" dynamics that are produced by various environmental factors and task-specific constraints.

---

43 See Bingham (1988, pp. 229-36) for a summary of the relevant two-way, three-way, and four way interactions between the different subsystems of the HAS.

Bingham distinguishes three broad classes of *incidental dynamics* (ibid., p. 239).[44] The first class includes all environmental factors which simply introduce novel inertial properties, dissipative properties, and mechanical constraints (e.g., when carrying a bucket on a handle, swinging open a door). The second class includes environmental resource potentials which are able to absorb, store, and return energy to the HAS (e.g., bouncing a ball, archery shooting). The third and most complex incidental dynamics are contributed by environmental resources which are independent sources of energy (e.g., operating a portable power saw, riding a horse, dancing with somebody).

According to Bingham, traditional approaches to the degrees of freedom-problem have been based on estimates in which the sums of individual parts of the HAS is considered in isolation from the specific task-environment in which a particular action has to be performed. This has frequently led to very large counts, which were based purely on the measurement of anatomical or mechanical properties of individual body parts. For instance, Kelso (1986; cited after Bingham, 1988, p. 259) simply refers to that fact that there are 100 joints and 792 muscles in the human body, or that there are over 40,000 muscles and tendons in the elephant's trunk, to illustrate the magnitude of the problem. Against this theoretically unconstrained way of counting, Bingham argues that there is no way of knowing how hard (in terms of movement dynamics) the problem really is for the HAS unless we can derive the relevant degrees of freedom from a detailed analysis of particular movement tasks in concrete environments.

To develop his alternative approach to the degrees of freedom problem, Bingham introduces the notion of a *task-specific device* (TSD). A TSD is "soft-assembled," i.e., of a temporary and dissoluble nature, and typically distributed across different anatomical structures. Conversely, one and the same anatomical structure can be employed in different TSDs. It is "assembled over the properties of both the organism and the environment" (ibid., p. 250), in the sense that it incorporates both organismic and environmental dynamic resources that are available to the HAS. Since the precise configuration of a soft-assembled TSD is dictated by task-specific goals and constraints, its dynamics are not identical to either the incidental dynamics or the inherent dynamics of the HAS (or any of its four subsystems). What allows psychologists to study TSDs as unified causal

---

44 Discussing Bingham's dynamicist framework, Wilson and Clark (2009, p. 65f) note that some of the main epistemic functions performed by our cognitive artifacts correspond roughly to the three kinds of incidental task dynamics identified by Bingham. These functions are (i) to provide otherwise inaccessible information and constraints (compass); (ii) to support off-loading and return of information (pen and paper); and (iii) to create new active sources of information-processing (word-processor).

wholes with the *lingua franca* of dynamical systems theory is that "the capacities required to mesh with and assemble incidental dynamics are the same as are required for the assembly of inherent dynamics alone" (ibid., pp. 238f). For instance, an analysis of the articulator (bones, joints…) dynamics of the HAS must track variations in inertia, since the inertia experienced at proximal joints varies continually with changes in the configuration of more distal joints. But the same principle applies to sources of variation which lie *outside* the body: for example, just wielding a hammer changes the inertial properties of the articulators in the inner link-segmental system.

What is the explanatory value of introducing TSDs? According to Bingham, it is expedient to analyze human actions first and foremost as a parade of dynamically coupled TSDs because the dynamics of a TSD are heavily *constrained* by the functional requirements and goals of particular tasks. This makes the TSD a more effectively controllable resource, because "controllability requires a reduction in the complexity of the resource dynamics […], [and] task specificity implies reduction in the dimensionality of the dynamics" (ibid., p. 241). Once we have a suitably precise analysis of the reduced TSD-dynamics, "[t]he challenge [for human movement science] is to work backwards from a description of the reduced dynamics to an understanding of the manner in which subsystem dynamics couple and co-constrain one another to produce the observed dynamical system" (ibid., p. 237). This pattern of reasoning is clearly an instance of what we have called the "top-down" method of working downwards. Following our tripartite template of an epistemic "business" cycle, the suggested methodology involves the following three steps:

1. [*investment*] The HAS selects a collection of organismic and environmental resources to assemble a TSD.
2. [*labor*] In virtue of the description as a TSD, the high-dimensional resource dynamics of the HAS is *reduced* to a much lower-dimensional dynamics.[45] This means in practice that the behavior of a TSD is predictable, and thus effectively controllable by a few parameters.
3. [*dividend*] A model of the reduced TSD-dynamics is used as a starting point to analyze the more complex, nonlinear component dynamics of the HAS.

Although our scheme is intended primarily as a rational reconstruction of Bingham's research methodology, we should not lose sight of the fact that it has real-world correspondences. As Bingham notes (ibid., p. 246, fn. 4), there is a rough analogy between the suggested procedure of analysis and the stage-like

45 In the next subsection, I shall say more about the notion of *reduction* that is at play here.

transitions displayed by the HAS when it learns to become properly dynamically coupled to an environmental resource in the context of performing a skilled action. For example, dynamical analyses of learning how to hula hoop (Balasubramaniam & Turvey, 2004), or of moving on a pedalo (Haken, 1996) have shown that the originally "sketchy" dynamic organization of the required movements is gradually refined as a result of learning, with the effect that the number of control degrees of freedom is greatly reduced until the entire coupled system is governed by a single complex order parameter (Haken, 1996). In addition, there is also a rough correspondence between Bingham's method of analysis and the initiation of skilled motor activity (Bingham, loc. cit.). However, this does certainly not imply that the three steps of analysis correspond to a computational sequence of discrete steps by which the brain controls the assembly process of the HAS, and thereby performs a reduction (cf. van Gelder, 1995).

3.4.2. Carving intelligent behavior at its joints

Building on our reconstruction, we now have to address three crucial points: (1) which notion of *reduction* is at play in Bingham's research strategy; (2) what are the epistemic dividends purchased by this reduction; and (3) what does all this imply for Rupert's deflationist understanding of dynamicist explanations? I will answer these questions in the indicated order.

Let's begin with the first question. The sense in which a higher-dimensional description of a dynamical system can be said to be *reduced* to a lower-dimension description refers to a certain type of mathematical procedure, in particular to numerical methods for simplifying a set of differential equations to the point where solutions can be found. A dynamical system is said to be *reduced* if its dynamics can be projected to a space of lower dimension without the loss of any (task-relevant) information (Strogatz, 1994). A simple example will provide an intuitive understanding of this procedure (cited after Bingham, 1988, pp. 246f). A widely used method of reduction is the use of constants of motion to "lock" the trajectories of a dynamical system into a region of its state space that can be projected to a lower-dimensional space. For instance, the evolution of a dynamical system along a closed-orbit trajectory on a (two-dimensional) surface can be described using two dimensions: the radius from the origin of the state space to the trajectory curve, and the angle formed by the radius. Now let's assume that the sum of the squares of position and velocity is kept a constant. This means that the trajectories lie on a circle in state space. For a circular trajectory, the radius is constant, which means that no information is lost if we describe the dynamics of the system using only one dimension, the angle. If this can be accomplished, we say that the former system, which has two degrees of freedom, has been mathematically reduced to the latter system, which has only a single degree of freedom.

Similarly, the instrumental value of positing TSDs is to reduce the high-dimensional resource dynamics of the HAS to a lower-dimensional description of a task-specific brain-body-environment system. It is precisely the promise of mathematical *reduction* which underpins the epistemic *irreducibility* of TSDs in Bingham's dynamicist analysis of human action. This sounds almost paradoxical, but that is only because the term "reduction" is used ambiguously here. The meaning of its mathematical use is practically the opposite of the relevant philosophical use of that term. By saying that some dynamic property of a complex system is *epistemically irreducible*, philosophers (unlike mathematicians) want to stress that those properties *cannot* be fully represented and/or predicted on the basis of our knowledge of the properties of the parts, their mode of combination, and the law-like regularities governing the behavior of the parts (cf. van Gulick, 2001; Stephan, 2006).

The two suggested meanings of *reduction* are, of course, systematically related. One way to see this connection is by way of Dennett's (1991b) conception of *real patterns*. For Dennett (ibid., p. 32), a stream of data has a *real pattern* if the information contained in the series is compressible, i.e., there is an effective way of representing the series that is more compact than enumerating each element of the series bit-by-bit. Such a compression is possible just in case the series is non-random. Based on this conception, a higher-level description is epistemically or "Dennett-irreducible" to a lower-level description just in case it reveals a real pattern. Accordingly, if we can mathematically *reduce* the high-dimensional intrinsic resource dynamics of the HAS to the lower-dimensional global dynamics of a TSD, we thereby show that a description of the latter (i.e., of the TSD) is Dennett-*irreducible* to a description of the former.

Moving to the second question, what do movement scientists have to gain from the combined outcomes of steps 1 and 2 of Bingham's methodology? Since the dynamics of the TSD are governed by much fewer parameters, it is a more effectively controllable resource. For a complex dynamical system to be effectively controllable, it must be stable and thus predictable. For one, it must be stable in the face of the necessary manipulations of its control parameters. In our example above, throwing a successful pass in football comes in a great variety of metric variations, e.g., depending on the distance to the receiver. Since the type and underlying structure of throwing motions remains invariant across these metric variations, it should be possible to achieve the required changes in action output smoothly by changing the values of its control parameters. Furthermore, the trajectories of an action system must be stable in the face of external perturbations and under variations of initial conditions. TSDs satisfy all these properties due to their reduced complexity. They allow for a much simpler, direct mapping between control parameters and target behavior, i.e., a one-to-one correspondence between particular parameters and specific types of task-related outputs (Bingham, 1988, p. 241).

As a result of the accomplished reduction in dimensionality, it subsequently becomes *easier* for movement scientists to work backwards from the TSD-dynamics to a characterization of the underlying component dynamics of the HAS (step 3). This may at first sound like a counterintuitive result – at least if one's intuitions are unreflectively biased in favor of familiar "bottom-up" decompositional explanations. Step 3 thus harks back to the original degrees-of-freedoms problem (Bingham, 1988, p. 247), which now involves solving an inverse problem in order to: first, reconstruct the structure and dimensionality of the pre-reduced dynamics which factors into the TSD; second, find out which particular dynamic properties from the total space of inherent resource dynamics are selected by the HAS for assembly (for a particular task).

Finally, but most importantly, what does all this imply for Rupert's deflationist interpretation of dynamicist explanations? Let us consider Bingham's framework as a test bed for the proper treatment of dynamicist models in psychology. Let's recall that for Rupert, cognitive psychology "hopes to explain the abilities of cognitive systems to interact with their environment. Cognitive psychologists would like to know how the organism acquires such abilities and how these abilities are manifest in different ways in different circumstances" (2009, p. 138). Note that we have also raised similar types of questions (about the HAS) in Bingham's framework, but the difference is that according to Bingham, these questions cannot be properly answered by means of traditional decomposition-and-localization strategies. Instead, they can only be addressed after the analysis of TSDs, i.e., in step 3 of the outlined procedure. The difference in viewpoint between "bottom-up" and "top-down" approaches to the behavior of complex systems is thus not primarily a dispute about what kinds of questions we can or should ask, but *how we should go about answering them.* The philosophical debate over how we should characterize the causal thicket of agent-environment interactions thus boils down to a dispute over scientific methodology. (One would hope this is progress!) It concerns the relative power and scope of decompositional versus dynamicist styles of explanation. If Beer, Bingham, and other proponents of a dynamicist stance are right, we must first figure out the autonomous dynamics of coupled brain-body-environment systems, and then successfully work downwards to model the nonautonomous dynamics of individual subsystem components. As we have seen, an important reason for analyzing the complex interactions between neural, bodily, and environmental resources as a single, coupled system is that the dynamics of the coupled system often yields a simpler, and more compact mathematical description than what we could have derived and/or predicted based on our knowledge of the parts. If the dynamicist stance is an empirically fruitful research program in psychology, the principle of CANDOR entails that we ought to take its ontological commitments at face value. Hence if dynamicist explanations essentially refer to entire brain-body-environment systems, I have argued that those systems are naturally understood as extended cognitive systems (Chapter 3.3.2).

So how should we properly frame the dispute between HEC and HEMC over the implications of dynamical systems-based models in psychology? Rupert seems to view the dispute primarily as a metaphysical issue, but I want to suggest that the debate is at the core an epistemological issue. Our difference in perspective is evident from Rupert's assessment of what it would take for the dynamicist argument in favor of the HEC to succeed. For Rupert, this would require that there is a substantial class of dynamicist models for which there is "no identifiable organismic state [which] has the appropriate causal structure to count as an internal realizer of the specifically organismic ability to use, or become coupled to, systems in the external environment" (2006, p. 21; see also ibid., 2009, pp. 146f). This is because the genuine lack of an internal realizer would undermine any attempt to attribute to the nervous system (or the organism) the particular cognitive capacity that is necessary (for a given task) to become coupled with the environment. However, according to Rupert's survey of dynamicist models, most of the currently available models of cognitive and sensorimotor processes do not support this claim, because in virtually every case, "the relevant mathematical models include separate quantities for organismic contributions and contributions of the external environment. This allows a clean separation between the organismically internal and the external contributions. Given this separability, there is a viable candidate for the internal realizer of the organismic capacity to couple with the environment" (ibid., p. 146). This would seem to leave the HEC-theorist with only the option to hunt for dynamicist models of intelligent behavior that invoke extended collective variables for which there is no tidy decoupling of neural, organismic, and environmental resources.

However, contrary to Rupert's perception of the debate, there is a more plausible epistemological view of the explanatory dividends of dynamicist models. The HEC-theorist need not deny that even in models of nonlinearly coupled dynamical systems we can typically still distinguish the *reference* of organismic variables from the *reference* of environmental variables (either taken as individual variables, or as relata in collective variables). For instance, in Beer's (2003) 16-dimensional model of the autonomous dynamics of the coupled CTRNN + agent + environment system, 14 variables clearly stand for the individual states of the 14 neurons of the CTRNN; the other two stand for spatial relationships between the agent and the environment. Beer's rationale for studying entire brain-body-environment systems is not that we can't distinguish or perhaps measure the individual states of different parts of the brain, the body, and the environment *at a given point in time*. His reasoning is rather that we cannot predict or explain how the dense web of continuous, concurrent, and interdependent *causal interactions between those states over time* leads to the generation of intelligent behavior unless we first analyze it in terms of a single, coupled dynamical system. In Beer's tripartite model, the explananda of layers 1 and 2 are *not* – contrary to what Rupert seems to suggest – the inherent ability *of* an organism to become coupled with its environment. That ability, and the organism-internal machinery on which it rests, is a separate explanandum – one that is most plausibly viewed as the target Beer's third layer of analysis. Consequently, to make Rupert's deflationist interpretation stick, he would not only have to show that there is "in principle" always a clean separation between the reference of organismic and environmental variables. That is the relatively easy part of his task. The much harder, but scientifically more relevant task would be to show that layer-1-type and layer-2-type dynamical analyses of the complex causal interactions between those variables are always *epistemically reducible* to layer-3-type analyses. Unless this has been demonstrated, the epistemic irreducibility of dynamicist models of entire brain-body-environment systems is sufficient to support our argument for the HEC.

3.5. Wrap-up

We began this chapter with a prelude about the show "cranium command," which nicely encapsulates the classical internalist view of the brain as the central executive manager of intelligent behavior. According to this view, the body is nothing but a passive recipient of instructions issued by the brain, the exclusive locus of cognitive control. My goal was to challenge this received view, by sketching an alternative picture of the intimate nexus between brain, body, and environment which is responsible for intelligent behavior. The emerging focus of this picture is much indebted to recent work in embodied robotics, biological cybernetics, computational neuroethology, and the growing use of dynamical systems theory in cognitive science. It has revealed more profound ways in which cognitive agents can be said to be *embodied*, and we examined a number of interesting implications of this richer notion of embodiment, especially concerning the thesis that cognition is *extended.*[46]

First, we saw the fundamental significance of ecological balance for the design of intelligent embodied agents. The *principle of ecological balance* refers to efficient ways in which embodied agents harness neural, bodily, and environmental resources so as to proficiently distribute the load of behavioral control. By reliably factoring the material, morphological, and dynamical properties of their bodies into the control task, embodied agents greatly simplify the task complexity for the neural system (while maintaining the same level of adaptive success). Second, we emphasized the epistemic use of *information self-generation* through sensorimotor coordination. Embodied agents strategically regulate their sensory access to the world by self-generated bodily movements – through touching, sucking, poking, grasping, pulling, pushing, gesturing, and other forms of exploratory behavior. These physical interactions with the world serve to generate "good" data, i.e., patterns of information which dovetail nicely with the morphology of their bodies, and the biologically indigeneous information-processing profiles of their brains. Finally, we challenged a central dogma of classical control theory, namely the theoretical separation of morphology and computation. Specifically, we saw how the morphology of an agent's body and its own movements can assume computational roles, and those *morphological computations* are functionally on a par with neural computations vis-à-vis controlling behavior. The resulting spread of computational/control load from the brain into the body or the world led us to propose the principle of *control-parity*:

(CONTROL-PARITY) If, as we confront a task, a part of BODY (WORLD) functions in a process which, were it to go on in BRAIN, we would accept as part of a process of behavioral control, then that part of BODY (WORLD) *is* (for that time) part of the control process.

The principle of control-parity is what proponents of embodied and situated cognition have in mind when they call for a reconceptualization of the "cranium command" view of behavioral control. Once it is established that behavioral *control* is extended, Clark and Chalmers' (1998) principle of parity (Chapter 2) further implies that *cognition* is extended, since the coordination and control of behavior is a widely accepted function of cognition.

The second part of the present chapter was concerned with the use of dynamical systems theory to study cognitive and sensorimotor processes as they

---

46 My discussion of embodied behavioral control in this chapter is greatly indebted to Andy Clark's lucid discussion of embodiment; cf. Clark (2008, esp. Chapters 1, 2, and 9.1).

unfold in real time. In particular, I discussed the multi-layered structure of Beer's (2003) dynamicist model of *minimally cognitive behavior*, and drew attention to Bingham's (1988) analysis of the organization of human action in terms of the dynamics of *task-specific devices*. I argued that the successful practice of dynamicist modeling supports the following claims: first, that the best explanation of (at least certain types or aspects of) intelligent behavior is likely to involve the dynamical analysis of (esp. nonlinearly) coupled brain-body-environment systems; second, that such an analysis is epistemically irreducible to other, decomposition-and-localization styles of explanation; third, that what makes a dynamical system cognitive is its capacity to generate certain types of intelligent behavior. Taken together, these essentially epistemological considerations suggest the following dynamicist argument for the HEC:

1. Understanding the dense, continuous, and often nonlinear causal interactions in which neural, bodily, and environmental factors conspire to generate intelligent behavior requires dynamicist explanations.
2. The success of dynamicist explanations requires that we treat the causal interactions among these factors as part of a single, coupled dynamical system.

C1: *Therefore*, understanding the complex ways in which neural, bodily, and environmental factors conspire to generate intelligent behavior requires that we treat them as part of a single, coupled dynamical system.

3. A dynamical system S is of cognitive type Φ just in case the temporal evolution of S is responsible for generating behavioral patterns of type Φ.

C2: *Therefore*, understanding the complex ways in which neural, bodily, and environmental factors conspire to generate intelligent behavior requires that we treat them as part of an extended cognitive system.

4. Language and writing as technologies of the mind

4.1. Introduction

When literacy first burst onto the academic scene through the ground-breaking works of Havelock (1963, 1976), McLuhan (1962), Goody and Watt (1963), Goody (1977), Ong (1982) and others, literate practices were considered not only as technologies of language, but also – and perhaps even more importantly – as technologies of the mind. In his famous lecture "Writing is a technology that restructures thought," Ong (1986) claimed that "without writing, the literate mind would not and could not think as it does, not only when engaged in writing but even when it is composing its thoughts in oral form" (ibid., p. 24). According to Ong, the most generalizable cognitive effect of writing is *diaeretic*, as he put it: writing is an instrument that divides and separates the word from the sound, the writer from the reader, the text from its interpretation, the embodied word from its natural surround, logic from rhetoric, knowledge from the knower, and being from time. But far from condemning writing for its distancing effects on human thought, Ong praised the virtues of writing as a cognitive technology that "is utterly invaluable and indeed essential for the realization of fuller, interior, human potentials." "Technology," he wrote, "properly interiorized, does not degrade human life but on the contrary enhances it" (ibid., p. 32).

But for every action, there is an equal and opposite reaction, and the history of literacy studies is no exception to that rule. Under closer scrutiny, many early generalizations about a "great divide" between oral and literate cultures, and thus between oral and literate minds, were shown to be boldly overstated, and in many cases had to be retracted (cf. Olson, 1991, 1994, 2008 for a review). First, to dub the advent of alphabetic writing as the pivotal breakthrough to "modernity" was seen by many critics as a thinly disguised celebration of Western culture as being superior to other, non-literate cultures. It carried the normative implication that "illiteracy" marks an undomesticated state of mind which ought to be overcome. Second, several putative features of the "literate" mentality turned out to be observed in non-literate cultures. Conversely, there were other, non-Western literate cultures which did not exhibit the forms of "logo-centric" speech and thought that had been associated with "modernity." Third, the predicted domain-general cognitive implications of literacy on the capacity for abstract, rational thinking were shown to be rather limited. Taken together, these results led many practitioners in the field of literacy studies to abandon the idea of a "literate mind" entirely, as a mischievous outpouring of a blatantly ethnocentric prejudice towards Western science.

My goal in this chapter is to breathe new life into the notion of a "literate mind,[47]" based on a conception of language, writing, and other representation systems as *cognitive artifacts*. The notion of a cognitive artifact, which is derived from ideas in cognitive science, human-computer interaction, cultural psychology, cognitive anthropology, and philosophy, has been developed within the framework of distributed cognition (Norman, 1988, 1991; Hutchins, 1995a; Perry, 2003; see also Chapter 1.2.3). Norman (1991) defines cognitive artifacts as "those artificial devices that maintain, display, or operate upon information in order to serve a representational function and that affect human cognitive performance" (ibid., p. 17). The renewed focus on the technological and social-cultural scaffolding in which people *learn* to think chimes well with older constructivist conceptions of the intricate relationship between language and thought. Central to these conceptions, which were pioneered by Soviet psychologists Vygotsky (1934/1986; Vygotsky & Cole, 1978) and Luria (1976), is the idea that language is not just an expedient medium for communication, but is an important vehicle of cognition. According to the hypothesis of extended cognition (HEC), language itself is a kind of "ultimate" cognitive artifact because of its potent capacity to re-shape the computational space that our brains must navigate in order to solve complex cognitive problems (Dennett, 1993, 1995, 2000; Jackendoff, 1996; Clark, 1998a, 2006a, 2008; Goody, 2000).

However, despite their shared underlying conception of language as a cognitive artifact, the main philosophical proponents of a tool-based view of language – notably Daniel Dennett and Andy Clark – have paid surprisingly little attention to the specific impact of *literacy* on language and cognition. Instead, they have typically focused on the similarities between spoken and written language as instances of "public symbol systems," as they frequently put it. It is tempting to think that this indiscriminate treatment is due to an implicit "written language bias" (Linell, 2005) that has haunted not only modern linguistics, but also (perhaps less perspicuously) cognitive science. This frequently unnoticed bias leads us to seriously underestimate the differences between our cognitive engagement with the spoken and written word. In a side note, Clark (1998, p. 182) once speculated that our use of spoken language may be as biologically proper to a human agent as the use of webs is to a spider – an "extended phenotype" in the

47 In this chapter, I am often using the notion of *literacy* in a broader sense, e.g., as including the competence of numeracy. This differs from its narrower use in literacy studies, where it is used almost exclusively to refer to the ability to read and write (using alphabetic characters).

sense of Dawkins (1982), whereas the use of written text may be more properly considered a genuine cognitive artifact. But these suggestions have not been pursued from a HEC-perspective as rigorously as they deserve.

Achieving a deeper understanding of the cognitive implications of literacy requires that we renew the intellectual bonds between cognitive psychology, linguistics, anthropology, cognitive neuroscience, philosophy, and literacy studies. As I see it, my part in this endeavor is mostly to serve as a "vector of memes" (Dawkins, 1976), in the hope that the tool-based conception of language will spread to new hosts in the field of literacy studies. Hence I will inevitably paint with a broad brush, leaving it for future work to elaborate the suggested notion of a literate mind with much greater sensitivity to the manifold uses of different types of writing systems by different types of individuals, social groups, and cultures.

The *leitmotif* of my attempted synthesis is a conception of literate minds as extended minds. Central to this agenda are ideas frequently dubbed as "constructivist," though the bewildering variety of views associated with this term makes it less than helpful as a label. For instance, in philosophy, constructivism is often understood as an opposition to ontological realism, implying e.g., that there is no fact of the matter whether certain statements are true or false, or that the existence of the external world is in some strong form dependent on the human mind. The psychological brand of constructivism that I shall advertize bears none of these implications. It concerns a thoroughly empirical thesis about the nature of the relationships between three types of representational systems – of language, thought, and writing systems. The spirit of my proposal has certain affinities to ideas that can be found in the writings of Vygotsky (1934/1986, Vygotsky & Cole, 1978), Luria (1976), and Piaget (1955), but the letter of my proposal is not bound to the specific content of any of their theories. The constructivist synthesis I shall try to forge hinges on my opposition to three popular views about the relationship between language and thought:

1. Artifacts are merely external aids for thinking.
2. Language is merely a vehicle for expressing thoughts
3. Writing is merely a transcription of speech

All three claims are familiar assertions of "*nothing-but*-ery." Statement (1) says that artifacts are nothing but products of, or input for real cognition which is done solely in the head. Statement (2) says that language is nothing but a conduit for transferring thoughts into and out of the mind, mostly for the purpose of interpersonal communication. Statement (3) says that our familiar writing systems are nothing but conventional public representations of speech in a graphical medium. While these statements certainly make distinct claims, I suspect that behind the widespread acceptance of all three views lurks a deeper-seated prejudice that has a strong grip on our thinking about the nature of mind. Roughly hewn, this prejudice can be stated as the assumption that the *representational re-description of information in a different format* does not deeply alter our cognitive abilities, but has merely practical implications. Let us call that thesis the assumption of "cognitive neutrality." An important goal of our constructivist inquiry is to spell out in greater detail what is meant by the assumption of cognitive neutrality, how it bears on the notion of a literate mind, and why we ought to reject it. This is, of course, a huge project which I can only hope to initiate in this chapter, by sneaking up on the issue of cognitive neutrality in three domains of distinctively human activities – the skillful use of cognitive tools in general, then language, and finally writing systems.

4.2. How cognitive artifacts work

Human beings have a special talent to modify the environments they inhabit through the creation of artifacts, and to transmit the cumulative products of these modifications to subsequent generations. A particularly interesting class of artifacts is what Norman (1991) has termed *cognitive artifacts* – "artificial devices that maintain, display, or operate upon information in order to serve a representational function and that affect human cognitive performance" (ibid., p. 17). Cognitive artifacts like the straightedge, the compass, pen and paper, writing systems, and word-processing software pervade our modern lives. But although we crucially rely on cognitive artifacts in so many of our intellectual endeavours, mainstream cognitive psychology has mostly focused on the cognitive functioning of biological brains that are unaided by technological and social-cultural resources. This neglect stems from an entrenched internalist view of the human mind that we have called "Cartesian materialism" (Chapter 1.1).

When we say that cognitive artifacts "make us smarter," what exactly do we mean by that? An obvious answer would seem to be that cognitive artifacts "amplify" our pre-existing cognitive capabilities. Just as a magnifying glass amplifies our visual capacities, Arabic numerals and a pocket calculator are said to amplify our arithmetical capacities. But as Norman (ibid.) points out, this answer is somewhat misleading, thus concealing the real source of their cognitive potency. He suggests that there are two different ways in which we can look at the role of artifacts: the *systemic view* and the *personal view* (Figure 10). From a systemic point of view, we consider the larger system which consists of a person who is interacting with an artifact to perform a cognitive task. Viewed from this perspective, it is often true that an artifact *amplifies* the performance of the system as a whole, compared to a person doing the same task without the aid of an artifact. But viewed from the perspective of the person who confronts a task through the artifact, it is more revealing to say that the artifact *transforms* the task at hand. Consider, for instance, how driving a car changes the task of locomotion for us. The skills and procedures that we have to learn in order to be able to drive a car are substantially different from our biologically more basic capacities to walk or to run.

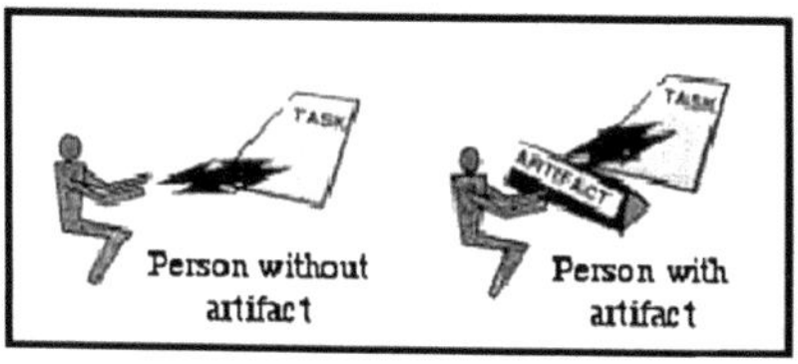

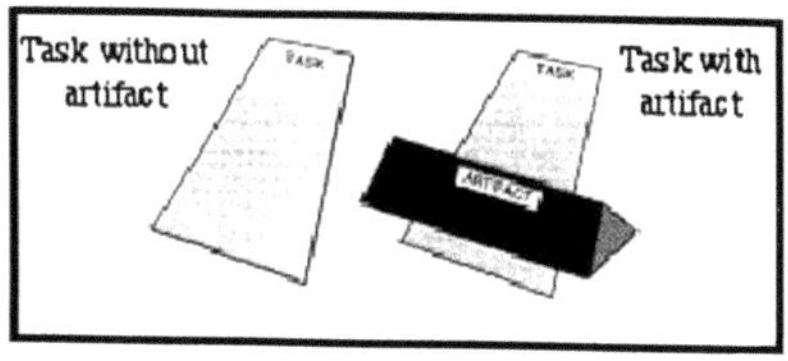

Figure 10 (from Norman, 1991). Left: from a systemic view (person + artifact), the artifact amplifies *task performance. Right: from a personal view, the artifact* transforms *the nature of the task she has to perform.*

Not all artifacts transform, of course. Some artifacts are indeed mere "amplifiers" insofar as they do not change the task which a user has to perform. For instance, a megaphone amplifies the intensity of one's voice such that it can be heard at much greater distances than would be otherwise possible. But the act of speaking remains essentially unchanged in form and content by the use of a megaphone. Now contrast the role of the megaphone – a non-cognitive artifact – with the function of our most highly prized cognitive artifacts. A classic example, taken from the realm of pre-GPS ship navigation, is the wide-spread use of a nautical slide rule for solving distance-rate-time problems, such as calculating a ship's speed in nautical miles per hour (Hutchins, 1995a, Chapter 3). Those are cognitively taxing tasks which navigators have to perform on a regular basis, with a high rate of accuracy, and sometimes under severe temporal and operational constraints. By using a slide rule, solving these problems becomes almost trivial. In a slide rule, multiplication and division are physically implemented as simple additions and subtractions of spatial displacements by means of two spatially juxtaposed logarithmic scales, which are already constructed in terms of the desired units as dictated by the problem at hand (Figure 11). Hence while the tasks facing the tool user lie in the perceptual-motor domain of scale-alignment operations, the computations that are achieved by the coupled system lie in the domain of arithmetic. Slide rules are well-designed cognitive artifacts, because our biological brains are more apt at manipulating physical objects in space and making perceptual judgments than at manipulating arbitrary strings of symbols (Norman, 2002).

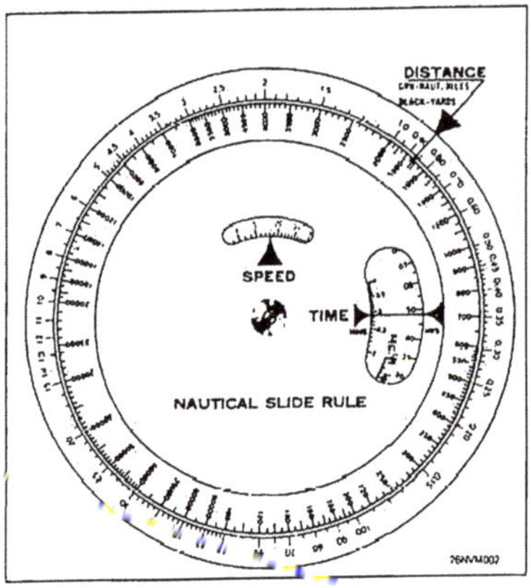

Figure 11. A nautical slide rule.[48]

4.3. Two views of language as a cognitive artifact

In much of contemporary cognitive science, the role of language as a cognitive artifact – an idea that figured prominently in the writings of Vygotsky and other proponents of the social-historical theory of cognitive development (Luria, 1979; Wertsch, 1985) – has been relatively neglected. Linguists, psychologists, and related fields have focused greatly on the more widely recognized role of language as an instrument for interpersonal communication. However, during the last few years, there has been a revival of "tool-based" approaches to language. Within the literature on distributed and/or extended cognition, the emerging "supra-communicative" (Clark, 1998) conception of language has been articulated in two ways, differing in their relative emphasis on what is considered to be the crucial role of language as a mind-tool. Associated with Daniel Dennett (e.g., 1991a, 1993, 1995, 2000) is the view that the main cognitive role of language is to sculpt and transform the psychological development of our brains. Let us call this the *internalization*-view of language. Associated with Andy Clark (e.g., 1997, 1998, 2002, 2006a, 2008) is the view that the main cognitive role of external linguistic symbols is to serve as perceptually stable, and easily manipulable material tokens which complement the more basic computational resources of our biological brains. Let us refer to the latter as the *complementarity*-view of language.

The internalization-view stresses the developmental effects by which our massive exposure to language impacts and partly re-configures our neural circuitry. At the center of this perspective, which has a distinctively Vygotskian out-

48 Image source: http://www.tpub.com/content/administration/14220/css/14220_211.htm.

look, is Dennett's portrayal of cognitive development as the internalization of culturally supplied resources that are first explored in a public arena. Most importantly, the advent of human language allowed our brains to be infected by *memes* – i.e., ideas which are encoded and thus can be culturally transmitted in the form of labels, words or sentences. During ontogeny, the bombardment of our brains with linguistic and other symbolic inputs significantly alters our higher-level cognitive architecture "courtesy of a myriad of microsettings in the plasticity of the brain" (Dennett, 1991a, p. 219). According to Dennett, the cortical re-organization results in the installation of what computer scientists call a new "virtual machine." A virtual machine is essentially a software duplicate of a physical machine that can also execute programs like its physical counterpart, but whose operations have no direct correspondence to the underlying hardware on which the virtual machine is implemented. Dennett has argued that the concept-using, conscious human mind which seems so familiar to us from a first-person point of view is itself a complex tool wrested from nature only through our prolonged immersion into a symbolic culture. Our approximately serial and compositionally structured virtual machine – the *Joycean machine* (Dennett, 1991a) – runs on a vast stream of language-like representations which are imperfectly implemented by the parallel distributed neural networks in our brains.[49] The strongly developmental emphasis of Dennett's internalization-view of language is closely related to the theory of neural constructivism (Quartz & Sejnowksi, 1997; Mareschall et al., 2007, 2008; cf. Chapter 1.3).

A different angle on the cognitive role of language has been pursued by Andy Clark. According to Clark's complementarity-view, the cognitive gains from our exposure to language do not primarily consist in any deep alterations of the basic information-processing profile of our biological brains. They rather stem from improving the functionality of our "same old" computational resources by re-deploying them in combination with new representational spaces. As a true cognitive artifact, the task-transforming contributions of language *for* the user do not depend on dramatic changes of the machinery *within* the user, but only on the user's ability to incorporate them properly into an essentially hybrid, environmentally extended problem-solving matrix. Central to Clark's vision is his rejection of the traditional view that linguistic structures do all their cognitive work by being exhaustively *translated* into some kind of content-matching inner code of thought, such as Fodor's (1975) symbolic expressions in "mentalese," or (on the other side of the spectrum) Churchland's (1989) high-dimensional, dis-

49 As the name suggests (it is named after James Joyce's "stream of consciousness"-technique of writing), Dennett (1991a) also links the emergence of human consciousness to the arrival of symbolic language.

tributed vector encodings in "neuralese." Although Fodor and Churchland obviously disagree about the computational structure of our inner cognitive machinery, they are both commited to a traditional *translation*-view of language. According to the translation-view, language is simply an expedient means for transmitting and exchanging information, a communicative tool whose concrete physical manifestation does not independently carry any cognitive significance over and above its content. As Clark (2002, p. 103) has pointed out, such a view is distinctly at odds with how we ordinarily think about tool use. For instance, the farmer doesn't exploit the functionality of his rake by first translating it into a kind of "muscle-ese," even though muscles are clearly involved to use the rake (ibid.). Similarly, Clark has argued that linguistic symbols do not stand in need of translation to serve as physically constitutive vehicles or substrates of cognitive processes.

To clarify his complementarity-view, Clark (2006a) distinguishes between three grades of "cognitive involvement" of language. On the lower end of involvement, Clark mentions the role of linguistic labels as perceptually stable targets for attention and learning. For instance, in a classic experiment, Thompson, Oden, & Boyse (1997) trained otherwise language-naïve chimpanzees to use different types of plastic chips to classify different kinds of visual stimuli (cf. Clark, 1998, for discussion). They found that only chimpanzees who had earlier learned to associate distinct plastic tokens with pair-wise relations of *sameness* or *difference* (e.g., cup-cup vs. cup-shoe) were later able to detect relations of higher-order sameness (e.g., between cup-shoe and cup-shoe pairs). Thompson et al. suggest that by mentally recalling the tags, the chimpanzees were effectively able to reduce the more abstract higher-order problem to a more lower-order perceptual problem. Another of Clark's examples is the development of rapid word learning in older children, which seems to require the ability to represent second-order, rule-like generalizations about the underlying categories to which a new word is likely to apply. For instance, knowing that the word "trap" stands for an artifact, a language learner can easily infer that the same word probably applies to similarly functioning things as well, regardless of whether they look the same, or are made of similar materials. Working from a connectionist perspective on language acquisition, Smith and Gasser (2005) have argued that the child's experience with arbitrary linguistic labels drives the construction of conceptual groupings that are based on more abstract commonalities and differences between objects.

A more deeply cognitive involvement of language concerns its role as a fulcrum for selectively maintaining attention that is directed at increasingly complex conjunctions of (esp. multimodal) stimuli. As Clark (ibid.) points out, a vivid illustration of this role comes from a study in which children were asked to retrieve an object whose location in a room was uniquely determinable only by remembering *both* the color of the adjacent wall and its geometry (Hermer-Vazquez, Spelke, & Katsnelson, 1999). Pre-linguistic infants were able to detect and use both types of cues in isolation, but apparently could not integrate the two cues. As a result of paying attention only to geometric features, they searched randomly in each of the two geometrically indiscernible sites. It was found that only children who were able to conjoin the relevant spatial and color terms in free speech were able to solve the complex task. Importantly, the same children failed to solve the same problem when they were simultaneously engaged in a speech-shadowing (i.e. repeating back speech) task, whereas their performance remained unaffected by a rhythm-shadowing task that was equally taxing for their working memory. These results suggest that children's linguistic abilities are actively involved in the cognitive integration of multi-modal information.

The third and highest grade of cognitive involvement implies that linguistic representations are constitutive parts of the mechanisms which underpin our thought processes. Following the influential account of mathematical knowledge by Stanislas Dehaene and colleagues (Dehaene & Cohen, 1995; Dehaene, 1997; Dehaene, Spelke, Pinel, Stanescu, & Tviskin, 1999; Dehaene, 2001), Clark's central example here is the complementary role which number words play for our arithmetical reasoning skills. For a better understanding of Clark's example, let me review some central features of Dehaene's intriguing account.

According to Dehaene et al.'s *triple-code* model, our adult knowledge of mathematics arises from the coordinated interplay of (at least) three distinct representational capacities (Figure 12). The first and biologically basic capacity, for which there are evolutionary precursors in non-human animals, is to form analogical representations of numerosity (e.g., *1-ness*, *2-ness*, *3-ness*, *more-than-that-ness*). Those representations can be understood as the distribution of activations along a mental "number line," which seems to underlie our approximative reasoning about magnitudes (e.g., numerosity discrimination, rough estimates) and has been shown to be language-independent. The neural mechanism for approximate reasoning involves the activation of bilateral areas of the parietal lobe that are also implicated in visual-spatial processing.

In contrast, our ability for exact arithmetic (e.g., calculating that $7 \times 6 = 42$) is not indigenous to the brain, but depends crucially on the acquisition of number words in a specific target language (e.g., "forty-two"). The neural mechanism for exact arithmetic relies on the activation of networks in the left temporal lobe that are also involved in word-association processes. This would explain, for instance, why children have to acquire the set of single-digit multiplication facts quite laboriously through verbal rote learning. Unlike for approximative addition, there is no "intuitive" algorithm on the number line available for exact multiplication.

Finally, acquiring more sophisticated forms of mathematical expertise appears to be essentially tied to the achievement of numeracy. For instance, being able to execute the familiar algorithms for multi-digit calculations requires that we can recognize, and know how to manipulate Hindu-Arabic numerals. The ability to perform such symbol-heavy calculations – usually only with the help of pen and paper – is enabled in the brain by a visual form-recognition system located in the occipito-temporal lobes of both hemispheres.

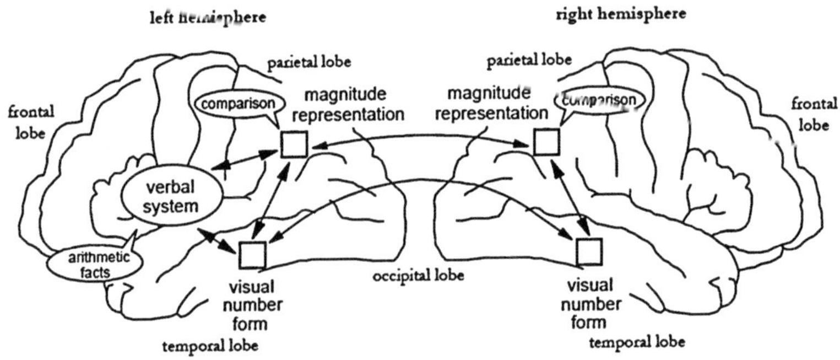

Figure 12 (from Dehaene & Cohen, 1995). Schematic functional and brain architecture of the triple-code model. Arrows indicate a functional transmission of information across different numerical codes, not an anatomically realistic depiction of neural pathways.

Converging evidence that there are distinct neural mechanisms for approximative and exact arithmetic comes from behavioral and brain-imaging data (Dehaene, 2001). First, brain lesions to the predicted areas show double dissociations between rote calculation and a conceptual understanding of quantities (Delazer & Benke, 1997), between exact calculation and approximation (Dehaene & Cohen, 1991), and even between multiplication and subtraction (Dehaene & Cohen, 1997). Second, exact mathematical knowledge acquired in a language-specific format cannot easily be transferred to the other (un-trained) language without considerable "switching costs" (due to the required internal translation), whereas approximate reasoning is language-independent and transfers without any switching cost (Dehaene et al., 1999). Third, brain imaging patterns during exact and approximate calculation confirm that distinct neural circuits are activated in each case (Dehaene et al., 1999).

When children begin to acquire a lexicon of number words and other elementary forms of verbally encoded arithmetic knowledge, they initially fail to appreciate what these symbols really stand for. For instance, children recognize that the word "three" stands for some number about a year before they have a sense of the precise numerical value to which it refers (Dehaene, 1997, p. 107). Later on, when children finally grasp that each number word refers to a distinct item in an unlimited series of exact quantities, this does not occur because the number word system has *re-configured* their basic analog "number sense" (e.g., adults still have no intuitive discrimination between 36-ness and 37-ness), or because number words have become *translated* into a symbolic medium of "pure" mathematical thought (cf. Clark, 2006a, pp. 8-10). Instead, it is because our brain manages to establish dense informational linkages between multiple, categorically distinct representational formats that "eventually leads, in numerate adults, to an integrated set of circuits that function with an appearance of non-modularity" (Dehaene, 2001, p. 12).

The active computational role which number words and numerals play for advanced mathematical cognition exemplifies the third and highest grade of cognitive involvement of language. According to Clark's complementarity-view of the relationship between language and thought, our manipulations of linguistic symbols tokened in some public code – e.g., as words in the air, as symbols printed on paper, or as shallow, perceptually grounded internal images thereof – are not merely "tools" for a purely inner organ of reason. They literally constitute proper parts of essentially hybrid thoughts.

However, the case of numerical cognition also casts new light on Clark and Chalmers' (1998) original formulation of the parity-principle (cf. Chapter 2). Parity trades on the assumption that cognitive processes carried out entirely inside the brain always provide the "golden standard" of cognition which sets the bar for any external processes that also ought to count as such. But if Dehaene is right about our development of numerical cognition, the suggested comparison with cognition-in-the-head surely puts the cart before the horse. Our adult mathematical competence emerges from the development of our numeracy skills and their functional integration with our intuitive, biologically basic number sense. Complex numerical cognition is thus a special type of cognitive process which is first and foremost carried out on paper, not in one's head (see Chapter 5.6 below for further discussion). There are rare cases of "mental calculators" with uncanny abilities of multiplying or factoring large numbers inside their heads. But as Dehaene (1997, Chapter 6) convincingly argues, these prodigious feats of mental calculation are almost certainly not the product of special innate talents (e.g., extra neural faculties dedicated to numerical knowledge). Most likely, they are simply the exceptional results of an unusually long and intense occupation with numerals which facilitates the discovery (and subsequent internalization) of nifty algorithms and heuristics that the rest of us would at least in principle also be capable of learning.[50] But at bottom, the same process of cognitive scaffolding is at work when we learn to use number words, number syntax, the counting sequence, finger-counting algorithms, the multiplication table, multi-digit calculation algorithms, and complex numerical concepts such as zero, negative numbers, and fractions. Viewed from the perspective of the "naked" brain, our adult mathematical minds are all calculating prodigies.

4.4. Departing from the assumption of cognitive neutrality

The time has come to pull together our previous considerations about language and distributed cognition, and use them to articulate a constructivist perspective from where we can better understand the cognitive implications of literacy. As we have said in the beginning, this perspective is based on a departure from the principle of "cognitive neutrality," i.e., the assumption that the representational re-description of information in different formats is always a cognitively neutral process. Let me now try to unpack this overarching theme into three related theses.

4.4.1. Representational re-descriptions

Our first thesis is that *cognitive artifacts enhance our cognitive abilities primarily by re-representing the tasks that our biologically evolved brains have to perform.* Herbert Simon (1981, p. 153) once said that "solving a problem simply means representing it so as to make the solution transparent." But as Hutchins (1995a, p. 117) added, what we mean by "transparent" depends on the properties of the person who is using those representations, and the context in which they are used. Let me portray the cognitive effects of Simson's representational re-description in three very different domains: (1) the use of specialized cognitive artifacts in the workplace, (2) the psycho-neural development of cognitive flexibility during human ontogeny, and (3) the active incorporation of language as a cognitive tool which straddles the boundary between the biological and the artificial.

My first example is Hutchins' (1995a) celebrated analysis of modern ship navigation. During the "fix cycle" – a navigational procedure performed to determine the current location of a ship – representations of the ship's relationship

---

50 Other spectacular examples of internalized cognitive scaffolds are the complex memory architectures developed by baroque monks and scholars (Carruthers, 1998), or the ability to play blindfold chess (Hearst & Knott, 2008).

to chosen objects in its environment ("landmarks") are created, transformed, and combined in ways that the solution to the original problem of position fixing eventually becomes transparent for the navigator. In the course of this transformation, information about physical space is propagated through a series of culturally constructed representational spaces that are built into various tools of the trade (e.g., the alidade-gyrocompass system, the bearing log, the hoey, and the nautical chart). Thanks to the social and technological distribution of cognitive labor, the navigator looking at his chart can tell with his "naked" eye that the position of the ship is where the plotted lines intersect. During the entire computation, information about the ship's position was implicitly contained in the sum of representational states that were propagated through various media. But it was not transparent to the navigator before it became graphically depicted on the chart.

My second example concerns the role of representational re-description in our cognitive development. In an attempt to synthesize nativist ("Fodorian") and constructivist ("Piagetian") perspectives on cognitive development, Karmiloff-Smith (1992) has argued that the internal organization of the mind/brain undergoes two complementary processes. On the one hand, there is a tendency towards the progressive specialization and modularization of cognitive processes into relatively autonomous domains (e.g., language, numbers, theory of the physical world, theory of the living world, theory of mind). But on the other hand, there is a concomitant trend towards the progressive explicitation and domain-general accessibility of previously implicit forms of knowledge. Offering an account of the second developmental trend, Karmiloff-Smith defends the hypothesis that the mind exploits information it has already stored, albeit in more procedural forms, by iteratively re-describing old information in new representational formats. In the course of this process, information that was formerly only represented *in* a cognitive system becomes available *to* that system (first within each domain, but then also across different domains), in forms that are easier to manipulate and thus more flexible in their use.

According to Karmiloff-Smith's model of *representational re-description*, the developing child appropriates and re-describes the mental representations which it constructs (in a given domain) on at least four different levels. Mental representations are first encoded in the implicit (I-level) form of special-purpose procedures for responding to stimuli in a given domain. They are computationally "bracketed," in the sense that no informational linkages within or across domains can yet be formed. This means that although the child is able to execute those domain-specific procedures, the information which is implicitly embedded in those procedures is not yet accessible for any other cognitive operations. The latter requires that the child re-describes these procedures on a more explicit

(E1) level, which makes the information they contain available to more sophisticated kinds of inferences. According to Karmillof-Smith, a child with an E1-level representational competence in a given domain is now able to induce systematic violations of her previously data-driven, veridical conception of that domain – e.g., violations which underlie the ability of older children to engage in pretend play, attribute false beliefs, and understand counterfactuals. Beyond level E-1, Karmiloff-Smith (1994) later introduced a bifurcation of the child's developmental path into two separate tracks, i.e., an explicit-2 (E2) versus explicit-3 (E3) level of representations. Both of these more advanced representational formats can be directly attained from E1. In E2, mental representations become available for conscious access, but not necessarily for verbal report (e.g., spatial, kinaesthetic experiences). In E3, mental representations are re-coded into an explicitly statable, linguistic format that makes information accessible and transferable across multiple cognitive domains.

As our third example, we have already discussed four roles of language as a cognitive artifact (cf. Clark, 1998). First, the use of linguistic labels to re-represent sensory stimuli facilitates the discovery of abstract patterns. Second, the performance of self-directed linguistic rehearsals helps to sculpt and drive our selective attention processes. Third, language provides a portable code that enables the transfer and integration of otherwise isolated domain-specific knowledge. Fourth, knowledge of language supports the development of otherwise unattainable forms of expertise (e.g., mathematical reasoning, symbolic logic). A fifth and hitherto unmentioned role of language, to which we shall now turn, is to support advanced forms of meta-cognition, i.e., our ability to reflect on, evaluate, and exercise voluntary control over our own thoughts.

4.4.2. "Frozen" thoughts

Our second thesis is that the ability to "freeze" thoughts in language enhances our meta-cognitive abilities of self-reflection, self-evaluation, and self-manipulation. The role of language as a vehicle of meta-cognition – i.e., a cluster of powerful techniques for thinking about thinking, which seems to be another distinctively human cognitive trait – has recently been explored by a number of different writers. Frankish (1998a, 1998b) has argued that the rehearsal of sentences in inner speech allows us to make deliberate decisions as to whether we should accept or reject certain propositions, based on a careful reflection on what we would commit ourselves to thinking or doing conditional on their acceptance. Jackendoff (1996) has suggested that the mental rehearsal of verbal formulations is the primary means by which our own thoughts become objects of attention and mental scrutiny. Unlike any other representational medium, language is uniquely apt to "fix" complex trains of thought as ordered sequences of symbols that we can inspect, pick apart, and critique. Similarly, Clark (1998) has emphasized the constitutive role of language for what he calls "second-order cognitive dynamics" (ibid., p. 177). Those include the ability to recognize and correct flaws in our plans or arguments, reflect on the unreliability of our own judgments under certain types of conditions, foresee situations in which we would be reliably disposed to form new beliefs, and then actively create those situations – i.e. the kinds of things scientists do when they plan their experiments (on this issue, see also Fodor, 1994, pp. 91-102).

Olson (2007) has argued that children's developing sense of responsibility, obligation, and self-control is dependent on language in several ways. First, children's ability to ascribe intentions to themselves, and to accept responsibility for their actions, is at least partly derived from parental verbalized efforts to hold their behavior accountable to certain rules and normative standards. It is during those interactions that children learn the necessary linguistic means to ascribe intentions to others, but also to themselves. Second, children's transition from being held accountable for their actions by others to the explicit self-ascription and avowal of their own intentions seems to be closely related to their comprehension of the linguistic device of quotation. In particular, Olson argues that the ability to repeat, rehearse, report, and generally quote speech enables a child to "decouple" the content from the attitude of intentions. For instance, when we promise to take on responsibility for our actions, we understand that a *promise* is an expression of an intention, not of an action that may (or may not) be the result of an intention. As Olson points out, younger children often conflate the making of a promise with the fulfillment of its outcome, saying things like, "it's not a promise because he didn't do it." En route to the self-ascription of intentions and the avowal of obligations which they imply, children must first become competent rule-followers. This requires that they can hold an utterance or rule in mind in the form of a quoted expression, and that they grasp the causal connection between their avowal of the rule and their actions.

Dennett (2000) has aptly characterized the human thinker as a master of "florid" representational practices. The mark of florid representing is that it is done deliberately, knowingly, and self-consciously, as opposed to the "pastel" forms of representations that we share with many other non-human animals. For instance, a classic example of a pastel mental representation would be the neural firings in the frog's brain which indicate the presence of a fly in its visual field (Lettvin et al., 1959). According to Dennett, there can be no florid representing without meta-representing, since the meta-cognitive awareness that is required for florid representing involves the competence to *appreciate that* what one is doing is an act of representing. The meta-representational capacities of thinking about thinking should also be distinguished from the biologically more basic capacity for having higher-order mental states, such as having beliefs about the beliefs of others (i.e., having a "theory of mind" in the sense of Premack & Woodruff, 1978, or Baron-Cohen, 1994). As Dennett (1995) put it – chimpanzees may have a basic theory of mind, but "they never get to compare notes. They never dispute over attributions, and ask for the grounds for each others' conclusions" (ibid., p. 380).

What allows human beings to engage in florid representational practices? As Dennett argues, it is not because nature has endowed us with upgraded version of "theory of mind" capacities, but because we are the species that talks. When we learn to speak, we first learn to use language in a more pastel fashion to refer to things. But in the course of linguistic development, children stumble upon the insight that words and sentences are a *sui generis* class of things that stand for other things. This epiphany opens the door for more "witting" uses of language, as can be observed in children's rhymes and wordplays. Over time, our florid experience with linguistic symbols helps to generate a new class of "thought objects" that are literally accessible in higher-order thoughts. Dennett's "outside-in" route to meta-cognition has a distinctively Vygotskian flavor, because it emphasizes the developmentally prior use of language for interpersonal communication. As a medium of communication, language provides a public code which is uniquely suited to present, inspect, and critically discuss ideas. Once the same code is internalized by the brain, and internally directed towards its own operations, it becomes a kind of "private" language that supports potent forms of meta-cognition.

4.4.3. Cognitive bootstrapping

Our third thesis is that *language is a powerful resource for "cognitive bootstrapping."* What does that mean? We have said that according to the perspective of extended cognition, the secret of our cognitive success is that "our brains make the world smart so that we can be dumb in peace" (Clark, 1997, p. 180). If this hypothesis is used to explain the evolutionary explosion of intelligence in the human lineage, it seemingly raises a concern that Clark (2001a, p. 150) termed the "paradox of active stupidity." Only *clever* brains, so the worry goes, would be able to design a smart world in the first place. In order to avoid a vicious regress, we'd thus have to look for other distinctive features of the brain (or the body) that must have made the crucial difference. But as proponents of the HEC have argued, this is not a real threat to the tool-centered hypothesis. The so-called regress never gets started in the first place because the initial discovery and subsequent evolution of successful tool use need not always require a process of deliberate design (cf. Dennett, 1995, 2000; Clark, 1997, 2001, 2003).

In order to illustrate the idea of "machines made from found objects," Dennett (2000) gives an interesting interpretation of Köhler's (1925) widely publicized studies of chimpanzees. Köhler famously found that chimpanzees were able to figure out how to use various tools in order to gain access to food (e.g., grabbing a banana that was out of reach by standing on a stack of boxes, or by inserting a narrow stick into a thicker one). Contrary to Köhler's enthusiasm over the chimps' genuine "insight" (*Einsicht*) into the corresponding "problem space," Dennett suggests that the animals were at first merely trying to familiarize themselves with objects in their environment. They were busy manipulating boxes and sticks as part of instinct-driven routines to tidy up their environments for action, rather than conceptualizing their environments *as* possible tools for action. Once they understood the basic mechanical properties of the objects they had found, some of them must have hit upon the "good trick" almost by chance, without systematically considering the environment as a possible means to their ends. It happened as a side-effect of exploratory behavior, combined with simple trial-and-error learning.

Dennett goes on to argue that language itself is perhaps the most impressive example of a "found object" that can be turned into an extremely powerful (virtual) machine. Language is reliably present in our environment. Like money or music, language is a mind-tool that was never created at once by any single individual. It accumulates structure as a result of being used, and modified through its use, by thousands of generations of speakers. But once language is installed in the human brain – a process which carries a relatively low cognitive load, since our maturing brains are naturally disposed to acquire language – it has a seemingly endless, and open-ended potential to be recruited for an unforeseeable variety of cognitive purposes.

4.5. Literacy, numeracy, and extended cognition

4.5.1. Clark's tacit "written language bias"

Despite viewing language as a potent cognitive tool, HEC-theorists have paid surprisingly little attention to the specific impact of *literacy* skills and technologies on the human mind. As a rare exception, Clark (1998, p. 182) once speculated that our use of spoken language may be as biologically proper to us as the use of webs is to the spider, whereas the use of written texts truly transcends the realm of the natural and may thus be more properly considered a genuine cognitive artifact. But the fact that such remarks are typically relegated to side notes suggests that the distinction between language and writing is not assumed to carry much weight. In their theoretical treatments, Clark and Dennett tend to focus on the similarities between spoken and written language, considering them pretty much interchangeably as instances of "public symbol systems." It is tempting to think that this indiscriminate treatment reveals an implicit "written language bias" (Linell, 2005) of the HEC. This bias may have led its main proponents to underestimate the differences among our cognitive engagements with the spoken and written word. In this section, my goal is to highlight some of these differences.

As characterized by Linell (2005), the *written language bias* (WLB) refers to the paradoxical situation in which the language sciences find themselves following the lead of Saussure, Bloomfield, Chomsky, and other founding figures of modern linguistics. On the one hand, there seems to be an almost unanimous agreement on the theoretical primacy of spoken language. Yet at the same time, language is described from conceptual and methodological vantage points that are historically derived from our literate practices, in particular from the normative concerns of cultivating, standardizing, and teaching written languages. According to Linell, the WLB explains the tendency of linguists, educational practitioners, and literate folk in general to treat properties that are really specific to written language as faithful representations of speech. In his book, Linell (ibid., Chapter 5) lists 101 features of what he calls the "WLB-complex." The WLB-complex includes the literate conception of language as a single, unitary phenomenon (e.g., a set of abstract objects that are governed by rules), the idea that sentences as expressions of propositional content are the fundamental constituents of language, and that speech has a discrete, segmental structure. Linell does not try to argue that there is no empirical basis at all in spoken language for any of these claims, or that they ought to be rejected in their entirety. Instead, his point is rather that the WLB exaggerates and thus selectively reinforces certain features of written language that are not equally representative of speech (ibid., pp. 35-36, 123-26). Similarly, I believe that Clark's conception of language as a cognitive artifact is heavily influenced by our deep familiarity with alphabetic writing systems. Let me mention a few symptoms of Clark's tacit WLB-bias.

Clark's (1998) discussion of the cognitive benefits accrued through linguistic methods of data manipulation and representation is most obviously indebted to our experience with written texts.[51] For example, Clark refers to the process of crafting complex arguments by means of "creating, putting aside, and reorganizing chunks of texts," his physical as well as intellectual interaction with "source texts and papers full of notes," and how "the real environment of printed words and symbols allows us to search, store, sequence, and reorganize data in ways alien to the on-board repertoire of the brain" (ibid., p. 173). Composition is a literate practice *par excellence*. Unlike ordinary spoken discourse, the process

51 This is even pointed out by Clark in the passage at hand.

of writing underscores the need to organize one's thoughts in a form that is suitable for expression within the limited resources of a public code. Olson (2001, p. 248) draws attention to the fact that "writing is never mere expression but rather involves editing and revising." As he points out, young children who first learn to write spend as much time erasing as actually writing (Graves, 1975). In a study comparing the oral and written productions of children, Hildyard and Hidi (1986) found that there were no significant differences until the age of twelve. After this point, the structural complexity of written texts began to take off, and this spurt began when children began to practice the art of revising texts. Horowitz and Olson (2007) have shown how students of writing first struggle to find their own "voice" in the process of citing, paraphrasing, and taking multiple perspectives.

In his explanation of what makes sentence-based linguistic rehearsal particularly well suited as a medium for meta-cognitive reflection, Clark (1998, p. 178) specifically refers to its properties as a code which "minimizes contextuality" (i.e., words retain more-or-less the same meaning in the different sentences in which they occur), is "effectively modality-neutral" (i.e., preserves sameness of content that is yielded by different sorts of perceptual input), and allows the easy rote memorization of simple strings. In similar vein, Clark (2006a, p. 11) praises language for its "arbitrary amodal nature, extreme compactness and abstraction, compositional structure, and so on," which he says complement the biologically basic computational resources of our brains. The suggested characterization of language fits the bill of Linell's WLB on several counts.

To begin with, the portrayal of language as an effectively amodal code carries the risk of distorting the richly multi-modal nature of our most elementary forms of talking to each other. Face-to-face conversations do not consist in exchanging amodal symbols between a "sender" and a "receiver" through a dedicated linguistic channel (i.e., the "text-messaging" model). The "situated order" (WLB #18) of dialogues is grounded in the social interactions between people who are dynamically coupled, in high bandwidth, through their actions, gestures, streams of talk, and a shared (and partly co-constructed) environment. Moreover, Clark's emphasis on *sentences* recapitulates the traditional view that sentences are syntactically and semantically integrated, isolable entities which form the basic units of linguistic processing (WLB #30). This reflects the Fregean assumption (WLB #79) that sentences are expressions of "complete thoughts" which have a propositional deep structure. The theoretical emphasis on sentences is prompted by our experience as writers, i.e., a process which forces us to verbalize the contents of our thoughts sentence by sentence. During dialogical interactions, on the other hand, our utterances are directly anchored in concrete conversational contexts. Hence the form of spoken discourse can afford to be deeply allusive, which often makes it implausible to analyze spoken language as a series of "incomplete" expressions of propositionally structured thoughts.

The putative context-independence of word meaning (WLB #46, 49) is also greatly exaggerated by the occurrence of words as printed symbols, for a variety of reasons. First, the classification of phonemes, words, and sentences as tokens of the same linguistic type is much less unequivocal if language is printed on paper (WLB #14). In the standardized form of a printed text, all tokens of the same type of word *look* exactly alike (if they are cast in the same font). Moreover, our rigid linguistic classifications are overlaid with norms that were designed by language cultivators and pedagogues with the primary purpose of teaching (lots of) people to write and read a standardized variant of written language. Hence these classifications are heavily influenced by dictionaries which define word types, and prescriptive grammars that define sentence types. The same classifications are much less applicable to spoken discourse, where situated word meanings (WLB #47, 58) and prosodic enactments (WLB #24, 25) are often situation-specific. Second, language on paper appears as a string of symbols that are spatially organized but lack a temporal dimension. This makes them available to be inspected in their entirety at one and the same time (WLB #16). But real-time speech is a dynamic and thus intrinsically transient phenomenon, always communicated and interpreted over time. Subtle aspects of timing and pacing speech are crucial aspects of successful linguistic communication (WLB #76). Third, the permanence of written texts as an array of repeatable, structured verbal objects suggests the idea that there is a singular and stable, *literal* (!) meaning of linguistic expressions (WLB #71). This is because texts can be re-read, and re-told over and over without any significant changes, whereas the activity of saying something is always contextually determined, and hence does not remain invariant across multiple instantiations. The level of verbal precision afforded by written language suggests the idea that the process of interpreting an utterance is essentially to recover a unique speaker intention which is determined by its written expression (WLB #78). But this conception of interpretation is most plausible only for a narrow range of erudite linguistic practices where exact wording is crucial. We often overlook that specialized forms of written discourse such as legal documents, administrative regulations, or mathematical notations have been specifically designed to support maximally form-based interpretations.

Let us pause here, and consider what I take to be the point of this brief exegetical exercise. Most importantly, my appetite for WLB-hunting should not be misconstrued as a dismissal of Clark's conception of language as a powerful cognitive artifact. Nor am I directly concerned with Linell's project, which is to overcome the WLB by promoting an alternative linguistic framework for studying talk-in-interaction as a dialogical, situated practice. My own goal is rather to combine Clark's conception of language with Linell's insightful diagnosis of the WLB. In particular, I want to suggest that those features of Clark's presentation which can be viewed as symptoms of an unwarranted written language *bias* (if they are predicated of language simpliciter) can be plausibly understood from a HEC-perspective as potential causes of a genuine written language *effect*. To demonstrate the intended change of perspective, let me revisit some of the previously outlined cognitive functions of language, but with our focus critically shifted to the cognitive impact of our literacy and numeracy skills.

4.5.2. Numeracy as representational re-description

Mathematical reasoning is widely regarded as a prime exemplar of an "abstract" and "purely symbolic" domain of human cognition. According to many cognitive-scientific accounts, our capacity for mathematical reasoning depends on the formal, rule-based manipulation of mental symbols (Anderson, 2005; Stylianou, 2002). The "amodal" representational format of mental symbols in one's head is assumed to be entirely independent of any visually distinctive features of mathematical notation on paper. While alternative "embodied" accounts of mathematical knowledge have stressed that many abstract mathematical concepts are processed through concrete, spatial metaphors (Lakoff & Nuñez, 2000), those accounts still contrast a physically "grounded" understanding of mathematics with the "ungrounded" computational process of crunching purely symbolic expressions. Challenging both types of accounts, Landy and Goldstone (2007) have argued that symbolic and non-symbolic forms of mathematical reasoning are equally grounded in perceptually based mental representations.

In their key study, Landy and Goldstone asked subjects to judge whether certain algebraic equalities were necessarily true. Performing such a relational reasoning task requires the correct application of operator precedence rules, e.g., the rule that multiplication precedes addition (Figure 13). Landy and Goldstone induced different perceptually based groupings of algebraic terms that would either support or interfere with our notational conventions. The tendency of subjects to group algebraic expressions in certain ways was manipulated by a number of perceptually salient but mathematically irrelevant factors. Those factors included perceptual *Gestalt* principles (e.g., physical proximity, connectedness, or common region), but also more complex patterns such as the format-based similarity of notational schemata, and even the proximity of letters in the alphabet. The goal of their experiment was to see whether inconsistent perceptual groupings could override subjects' (independently tested) knowledge of precedence rules. Although the subjects were fully aware that perceptual groupings should not play a role for their judgment, their accuracy was much higher when the induced groupings were consistent with the true order of precedence rules. Moreover, the observed increase in accuracy was significantly greater when the correct judgment was crucially dependent on the order of operations.

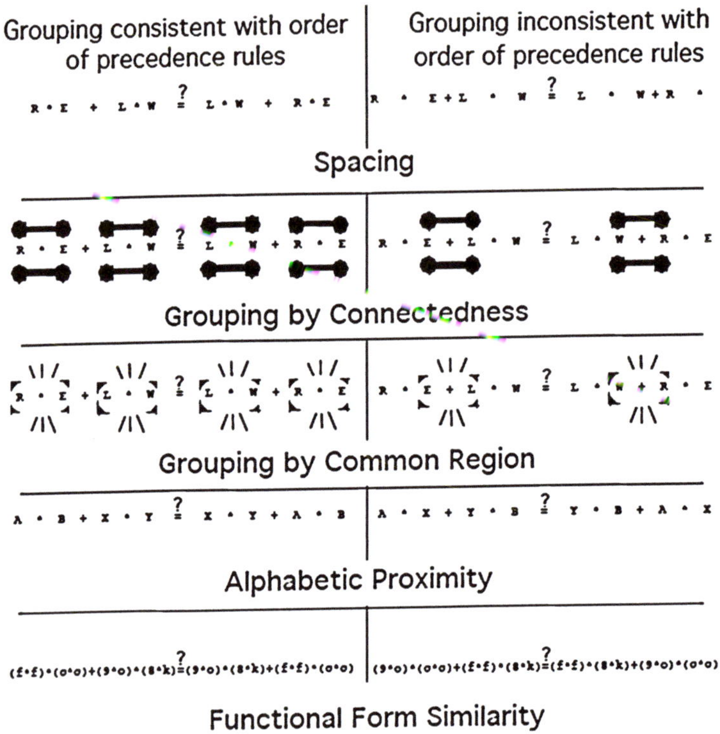

Figure 13 (from Goldstone, Landy, & Ji, 2008). Sample stimuli from several experiments on the impact of perceptual groupings in an equation-verification task (Landy & Goldstone, 2007). See the text for further explanation.

Landy and Goldstone suggest that these effects demonstrate a genuine *cognitive* illusion in the domain of mathematics. In particular, if they were simply due to ordinary perceptual errors (e.g., to decipher a symbol, or to find matching terms from each side), or a failure to remember whether a certain item has already been checked, one would expect that errors should occur with the *same* frequency regardless of whether the correct judgment was sensitive to the order of operations or not. But this was clearly not the case. The data show that perceptual groupings hurt accuracy only in the sensitive condition. This implies that the nature of perceptual groupings directly affects the underlying representational format in which people process the relational structure of algebraic expressions.

As Landy and Goldstone point out (ibid., p. 731), these results do not sit well with "two-stage" accounts of mathematical reasoning that assume a strict separation between the perception of external notations and proper mathematical cognition. According to such accounts, the visually distinctive features of the notation (such as the size, color, or non-ordinal position of graphemes) are merely used by the perceptual system as cues to categorize the written input tokens into discrete types. In the course of this process, the visual percepts of external symbols are "transduced" into a structured sequence of formal symbols that only retain information about their categorical status. Mathematical reasoning consists in the syntactic manipulation of amodal symbols in ways that reflect our knowledge of operator precedence rules. More generally speaking, we can say that two-stage accounts of mathematical reasoning are committed to a *translation*-view of the relationship between language and thought (Section 4.3).

In order to explain the observed effects of perceptual groupings, Landy and Goldstone propose a more parsimonious "interactive" account in which higher-level processes of "symbolic" reasoning are directly performed on perceptually grounded, image-like representations of the notation (ibid., p. 731). Perceptual processes are used to group ("chunk") notational stimuli at various constituent levels, which are then submitted to a visually based comparison process to verify whether they are the same. The comparison starts at the highest possible level, i.e., the level on which each side of the equation is chunked as a single item. If those items don't match, the comparison descends down the constituent hierarchy simply by adjusting the relevant spatial grouping, and then successively inspecting smaller and smaller visual chunks. Assuming that perceptual groupings define the default order in which the order of mathematical operations is evaluated, Landy and Goldstone's account nicely explains why the presence of inconsistent groupings would interfere with the visually based comparison process.

4.5.3. Literacy as representational re-description

The marginalization of writing in modern linguistics can be traced back to its founding fathers, Ferdinand de Saussure (1916/1983) and Leonard Bloomfield (1933). Both helped to perpetrate the widely held myth that writing is merely an ancillary device for the transcription of speech. For instance, Bloomfield said that "writing is not language, but merely a way of recording language by means of visible marks" (ibid., p. 21; cited after Olson, 1996, p. 156). Let us call this view the "copy" theory of writing. Its influence effectively banned the study of literacy from linguistics for the better part of the 20th century. It continues to persist because of a popular, albeit fallacious appeal to the earlier mentioned "paradox of active stupidity" (Chapter 4.4.3).

Consider, for example, the representation of sounds as discrete letters in an alphabetic script like ours. For a child to learn how to represent sounds as letters, or so it is argued, it must be able to parse speech into the relevant inventory of phonemic units which the script represents. This means that the child must already bring to the task of learning how to read or write a prior knowledge of the phonemic structure of language (*mutatis mutandis* for all other linguistic categories such as words, sentences, etc.). Consequently, writing must be a mere transcription of what is already known (i.e., the "assumption of cognitive neutrality").

Despite its popularity, the copy-theory of writing faces serious empirical difficulties (cf. Olson, 2002, pp. 158-60). First, children assume at first that written signs directly stand for objects and events in the world, rather than for words or sentences about those objects (Ferreiro 1991, 1994; Homer & Olson, 1999). Prereading children who were dictated words (such as "cat") which they had to write down spontaneously invented their own marks, often in form of a pseudostring. Now when they were asked to write "two cats," they simply repeated the original string twice, and similarly for "three cats." Second, in the corresponding reading task, pre-literate children were asked to interpret written phrases (e.g., "three little pigs") that were read to them aloud. When the last word of the three-word phrase was covered, children usually answered that the remaining string meant "two little pigs." Olson (2002, pp. 160f) cites a related study by Alice Moro in which the same response pattern was also observed among illiterate adults. Third, when children were asked to write down "no cats," many of the pre-readers replied that this could not be done. They evidently did not see how written marks could be used to denote the absence of things, even though they were perfectly capable of understanding the corresponding verbal expression "no cats." Fourth, Olson cites the historical development of event-based or object-based scripts prior to the invention of phonetic writing systems, which contradicts the order one might expect if the copy-theory were the psychologically most natural way of representing speech.

As an alternative to the copy-theory, David Olson (e.g., 1991, 1994, 2001, 2002, 2008) has extensively defended the constructivist view that writing severely transforms our conceptual and cognitive engagement with language. Olson's way out of the suggested "paradox of active stupidity" is similar to the response proposed earlier (Olson, 2002, pp. 156-58). On the one hand, Olson readily concedes that "it is plausible to suggest that to learn a notational system requires the metarepresentational concepts to understand how such symbols work" (ibid., p. 157). In particular, in order to be able to read and write, a child must learn how to parse one's speech in ways that are *compatible* with the units of the alphabetic script (e.g., letters, words, sentences). However, this does not require that the child must *already* be explicitly or even tacitly aware of speech *as* composed of discrete letters, words, and sentences, since that knowledge, too, can be acquired in the process of becoming literate (ibid., p. 156). As an illustration of Olson's point, consider what it takes to learn how to read a city map. A competent reader of city maps has to be able to understand that the reference grid in the map defines points in a culturally constructed space that is used to identify features of the townscape in physical space (e.g., locations, distances, etc.) But it would be wrong to infer that prior to a person's exposure to maps she must have already "implicitly" perceived the townscape in terms of a grid structure.

According to Olson, the single most profound implication of literacy is that it turns language into an *object* of reflection and meta-linguistic awareness. The highly supervised process of attaining literacy is uniquely apt to spark a learner's meta-linguistic knowledge, much more so than the unsupervised acquisition of speech. One could say (echoing Dennett) that unlike speech, writing is an intrinsically "florid" (Dennett, 2000) representational practice (Chapter 4.4.2). A student of reading/writing is inevitably forced to reflect on how certain aspects of linguistic structure ought to be represented as written marks, according to the specific categories afforded by the writing system. For instance, our familiar alphabetic script partly induces, and partly reinforces one's meta-linguistic awareness of the distinctions between sounds as represented by letters, the distinctions between words (in a sequence) as represented by spaces, the distinctions between clauses as represented by punctuation, between thematic elements as represented by topic sentences and paragraphs, and between types of discourse as represented by different genres (Olson, 1991, p. 263). A heightened sense of meta-linguistic awareness is known to be crucial for a wide range of advanced language skills (Olson, 2002, pp. 162f). Those include the ability to closely monitor one's choice of words, tense, aspect, and clause structure; to try to honor grammatical norms dictated by written standards; to adhere to a strict demarcation between sentence meaning and speaker meaning (Olson & Torrance, 1987); and to understand the meta-linguistic mechanism of quotation (Olson, 2001). Finally, the advent of writing most likely acted as a "seed" (Clark, 2001a, p. 151) which triggered a host of related linguistic, psychological, and social competencies. For instance, specialized forms of discourse (e.g., written prose) historically arose from the fact that members of a literate community were able to address each other in written form (Godzich & Kittay, 1987). Dealing with these novel forms of discourse, in turn, fostered the development of specialized intellectual competencies (Olson, 1991, p. 267; see also Olson, 1994).

It is important not to misconstrue the nature of Olson's constructivist literacy hypothesis – lest it becomes implausibly strong. His claim that in the process of attaining literacy, speakers learn to re-describe language along the categories provided by their script does not imply that those categories have *no* empirical basis at all in the structure of speech (Olson, 2002, p. 164; see also Linell, 2005, p. 125). For instance, there are clearly significant aspects of lexical structure that lend themselves to be analyzed in terms of words, and sounds have an internal phonological structure that lends itself to a discrete segmental analysis. In short, the transformative process of representational re-description should not be confused with the arbitrary invention of linguistic categories *ab initio*. The distinction between constructivism and creationism surely ought to be borne in mind. Olson's point is rather that our knowledge of language is partly co-constructed by the internalization of the very script that we learn to use for representing speech.

Challenging Olson's "meta-linguistic" literacy hypothesis, Karmiloff-Smith et al. (1996) have argued that children already possess a fairly rich meta-linguistic understanding of the concept of *word* that precedes literacy. To elicit children's meta-linguistic knowledge, they devised a nifty experimental technique that was specifically designed to engage children's normal routines for on-line linguistic processing, but momentarily take them off-line. During a typical trial, children would listen to a story which the narrator deliberately interrupted at certain points. The narrator then asked the children to repeat either "the last *word*" or "the last *thing*" that she read. Children as early as 4.5-5 years old were almost flawless in repeating the correct *words*, and not only for open class words (e.g., "ball"), but also for closed class words (e.g., "the"). The authors interpret their data in terms of Karmiloff-Smith's psycho-developmental model (1992) of representational re-description (Chapter 4.4.1). According to Karmiloff-Smith et al., 3-year olds have only tacit, I-level behavioral mastery of language, 4-5 year olds have advanced to an E-1 level of linguistic understanding, and 6-7 year olds have reached a fully explicit level of meta-linguistic awareness that can be re-deployed for off-line linguistic tasks (e.g., calls for word definitions, word counting, word-by-word dictation).

Does this show, as Karmiloff-Smith et al. suggest, that the development of meta-linguistic awareness is primarily an *endogenously* driven process and thus a precondition of literacy, rather than a largely *exogeneously* driven product of literate education, as claimed by Olson? In his defense, Olson has always been very careful to point out that writing is not the only means for turning language into an object of reflection and discourse (e.g., Olson, 1991, p. 264). First, children's rehearsal of rhymes, verses, and puns is also prone to raise their meta-linguistic awareness of sounds, though not necessarily in the same form that is dictated by an alphabetic script. Second, cross-linguistic studies have shown that every natural language contains at least a basic fragment of meta-linguistic expressions such as "tell," "say," "ask," or "lie" (Leech, 1983). Third, Olson has never claimed that internalizing the categories of a public script is simply a matter of "swallowing" external squiggles. It should be viewed as a constructive learning process in which the child attempts to integrate the categories of the script with her pre-existing knowledge of spoken language (2002, p. 157).

Moreover, the suggested cultural transformation of our linguistic knowledge though literate practices certainly goes well beyond the endogeneously driven developmental stages posited by Karmiloff-Smith's model. As an interesting example of the Vygotskian "outside-in" route of influence, Olson (1994, Chapters 6-7) has argued that the cultural spread of reading to a wider audience opened up a whole new world of interpretative discourse to deal with the problem of *literal meaning*, i.e., the hermeneutical problem of recovering an underlying, audience-directed speaker intention from a written text. This hitherto unknown form of interpretive discourse called for a more elaborate repertoire of verbs expressing speech acts and mental states, such as the need to distinguish expressions such as "assert" from "imply" (instead of merely using "say"); or between "assume," "infer," and "hypothesize" (rather than just saying "think"). In their analysis of the massive borrowing of vocabulary from Latin into English that occurred in the 16th and 17th century, Olson and Astington (1990) showed that a substantial part of this influx addressed precisely those emerging vocabulary needs. Again, their analysis should not be taken to imply that illiterate speakers cannot assume or infer things. However, having a rich corpus of fine-grained meta-representational terms is bound to sharpen our meta-cognitive ability to recognize and evaluate an assumption *as* an assumption, an inference *as* an inference, or a mental state *as* a hypothesis according to rigorous epistemological standards that are unavailable to the illiterate mind (Olson, 1994, pp. 275-82).

Similarly, as Linell (2005, Chapter 6) helpfully reminds us, the conceptions of language which come to us in the form of modern linguistics, language pedagogues, school teachers, and textbooks originally arose from political as well as practical tasks concerned with the spread of literate practices. These concerns included the establishment of scribal practices, the elaboration of standardized national languages, and of crafting a meta-linguistic vocabulary suitable for describing languages so that people could be efficiently taught to read and write. The emergence of literate institutions for producing and interpreting texts (e.g., church, courts, governments), educational institutions for inducing learners into the world on paper, and new media for technologizing the written word (e.g., printing press, computer, email) have continually affected the intellectual competences that are required to participate in a literate tradition (Olson, 1991, pp 252-54).

In sum, my own take on this debate is that the two seemingly competing constructivist accounts of representational re-description – the "Piagetian" inside-out route and the "Vygotskian" outside-in route – should be viewed as complementary rather than strictly incompatible.[52] From the standpoint of the deeply hybrid HEC-perspective, it seems eminently plausible that the representational re-description of language through externally driven loops first builds on, but eventually outstrips and thereby redirects the representational re-description of language through internally driven loops (Chapter 1.3).

4.5.4. A short "bootstrapping" history of writing

As a demonstration of how to overcome the "paradox of active stupidity," the cultural evolution of writing systems is rife with examples of "cognitive bootstrapping" (Chapter 4.4.3). A particularly striking illustration is Denise Schmandt-Besserat's (1992, 1996) intriguing account of the origin of the Mesopotamic cuneiform script, a precursor to our own alphabet. Her account directly contradicts the orthodox "pictographic" view, which is deeply rooted in the history of Western thought (Harris, 1986), that the cuneiform script began its life as a system of pictorial signs that were later redeployed as letters as their representational content became more abstract. Against this view, Schmandt-Besserat has provided compelling archeological evidence that the cuneiform script is in fact a direct descendant of a counting system of clay tokens that evolved as a ancient bookkeeping device in the prehistoric Near East, dating back into the ninth century BC. Schmandt-Besserat's multi-stage account of this historical trajectory is complex, and involves several surprising twists and turns.

During the first four thousand years of their use, "plain" tokens with basic geometrical shapes (such as cones, spheres, and disks) were used mostly for counting and accounting for agricultural goods. Around 3500 BC, the number of "complex" types and subtypes of tokens began to proliferate in the administrative centers of the South, where it became necessary to represent the increasing variety of goods that were manufactured in the big cities. These complex tokens

52 Olson (2002) contains a brief reply to Karmiloff-Smith's work (see esp. p. 157, p. 162). In response to a paper by Clark and Karmiloff-Smith (1993), Dennett (1993) has defended a somewhat more Vygotskian line. He argues that the general capacity of a child to re-describe its own internal cognitive states ("internal labeling") depends on her developing capacity to master and use a public language to label external stimuli ("external labeling"), which begins with the process of representing auditory inputs as sounds.

displayed a larger repertoire of geometrical forms, the addition of more naturalistic shapes, and a profusion of markings.

According to Schmandt-Besserat (1996, Chapter 5), the archaic token system was the earliest man-made public "code" – and thus a historical milestone in its form and content. First, the counting system was very easy to create and manipulate. Clay was the perfect material. It was commonly available, and had all the required properties: malleable at first, but convertible to the more durable ceramic. The basic types of shapes that were first in use were all easy to produce. Second, as a system of *concrete counting,* the token system superseded older systems of counting that were based on *tallies* (e.g., notched bones, pebbles). The new format of movable units was probably inspired by the earlier use of pebbles. Unlike notches, tokens can be flexibly arranged into groups of variable sizes and composition, and the results of those manipulations can be preserved. The main difference to *abstract counting* systems, which were only invented later, was that the quantity of items was not yet fully dissociated from their categorical content. Different shapes of tokens would thus be used to denote such varied things as discrete units of grain, measurements of land, or types of animals. However, the added principle of *object-specificity* meant that each token belonged to exactly one type, and every occurrence of the same type of token carried the same categorical meaning. For instance, a plain cone always stood for the same unit of grain. The number of items pertaining to a given category would still be marked with the token-iterative principle of the tally. For instance, three sheep would be represented as the equivalent of "sheep, sheep, sheep." The *additive* representational format of the tally makes it easy to compare quantities since the length of a tally bears an icnomic relationship to the numerical value which it represents (Norman, 1991). Finally, the graphical nature of the system made it independent of phonetics, so there were no language barriers in teaching it. It is therefore easy to imagine how the archaic token system must have swept through the Near East, especially in light of the economic needs for accounting created by the "Neolithic revolution" in agriculture.

The archaic token system was an impressive cognitive artifact that supported new forms of data storage and manipulation (Schmandt-Besserat, 1996, p. 96). First, it was the first mnemonic device that could store a potentially unlimited amount of data. Second, it brought more flexible ways of adding, subtracting, and manipulating data. Finally, it had a visually distinctive "user interface" which significantly enhanced the rational decision-making skills of the accountants who were put in charge of managing large amounts of economic data.

Early in the fourth millennium BC, the introduction of new methods of storing tokens in archives triggered a process that eventually transformed the archaic counting system in the most unexpected way. These new methods were necessary to ensure that collections of tokens representing a single account could be reliably identified and securely kept together (Figure 14). One popular method was to perforate tokens with a string which could be tied together and sealed off by solid lumps of clay (*bullae*) that would bear the mark of the economic transaction. Another method was to enclose the token counters in hollow clay balls (*bulla-envelopes*). Envelopes were preferably used for plain tokens, but there was apparently a cross-over between the two methods. A distinctive drawback of using envelopes is that they conceal the tokens which are kept inside, thus making it impossible to verify the contents of an account without breaking its enclosure.

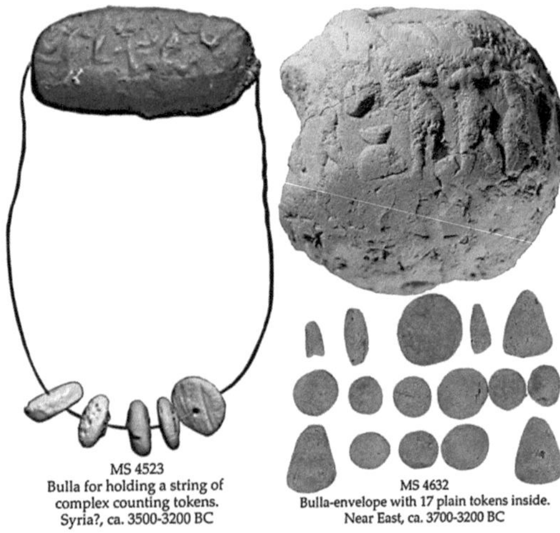

Figure 14 (Image source: http://www.schoyencollection.com/math). Two different methods of storing clay counters that were used in the ancient Near East for accounting purposes (Fourth millennium BC)

As a result, accountants began to develop new techniques for representing the contents of the bulla-envelopes on the outside. Of the many techniques that were tried out, only two methods survived. One was to stamp the tokens into the soft clay of the envelope so they would leave an impression on its outer surface before they were stored away inside, thus preserving a one-to-one correspondence to their quantity and shape. The other method was to inscribe markings on the outside surface by means of a sharp stick (*stylus*). Initially, these impressed markings were merely indicators of the actual counters contained within the envelope. But at some point, some clever accountants – perhaps not unlike Köhler's chimpanzees (Chapter 4.4.3) – must have realized that the tokens inside were made redundant by the presence of the outer markings. Eventually, the ancient practice of keeping the physical tokens dropped out of the picture, and was replaced by just impressing the corresponding signs into solid clay tablets. These signs perpetuated the shape of the older tokens, but no longer encoded a message that was contained elsewhere – the signs *became* the message (ibid., p. 55). This transition, which was born out of the necessities of archival storage practices, acted as a "seed" that led to a veritable revolution in writing technologies.

From this point on, a dichotomy of scripts developed. Wedges and circular-shaped signs continued to be produced by means of impression, whereas oval and triangular signs were typically incised into tablets with a stylus. According to Schmandt-Besserat (ibid., pp. 68-79), the divergent evolution of scripts can be explained by the fact that their three-dimensional ancestors – plain versus complex tokens – were used by different people for different purposes and stored in different manners. Complex tokens with all their sophisticated incisions, whose subtypes multiplied in response to the rapid increase of manufactured goods in the cities, did not lend themselves to being impressed. A legible, unambiguous two-dimensional representation of their distinctive three-dimensional shapes required a degree of accuracy that was only supported by the incision method, which began to be performed with the help of a specialized triangular stylus. The system of incised signs on clay tablets that developed out of this technique was the latest stage in the evolution of writing before it finally turned into the Sumeric cuneiform script. The clay tablets continued to be made of the same material as the older envelopes, and also had a similar shape and size. Moreover, there is a direct continuity in the shapes of the incised symbols, and their spatial organization along horizontally structured lines. But the inventory of signs grew quickly as scribes began to exploit the affordances of writing on a two-dimensional surface.

As a way of summing up her account, Schmandt-Besserat (ibid., p. 98) argues that the idea of writing little "pictures" of clay tokens instead of storing the tokens as such brought four major innovations to our capacities for storing and processing information. First, compared to the transient collections of clay tokens, symbols written on a tablet are much easier to preserve permanently. Second, tablets accommodate the representation of more diversified information by making it possible to assign specific parts of the two-dimensional array to record particular types of data. This led to the development of listing conventions, such as the custom of putting the symbols for the sponsor or recipient of an economic transaction underneath the symbols for the exchanged goods. Third, it led to the evolution of abstract counting systems in which information about the number of items is strictly separated from their category. For instance, in order to represent three sheep, one would use two separate types of symbols: one for the category sheep, another one for the quantity three. Harris (1986, pp. 145f) has argued that the step from token-iterative to "type-slotting" systems for recording numerical information marks one of the great inventions in the evolution of writing systems. The slotting principle necessitates a primitive syntax, because the interpretation of the complex symbolic expression is now essentially mediated by the ways in which their component parts are meant to be combined. Fourth, the "generative turn" of the pictographic system also paved the way for writing to become phonetic. As Olson (1994, p. 74) points out, a script with a syntax provided, for the first time in history, a suitable model for speech. It was now finally possible to conceive of marks as representing *words* rather than things.

In sum, Schmandt-Besserat's account reveals thath there is a grain of truth in the popular pictographic story. It is true that the signs which were first incised on tablets were "pictures" in some sense – but not, as traditionally conceived, of objects in the world, but rather of clay artifacts that had previously been used as an archaic counting device. If Schmandt-Besserat is right, writing really emerged from the administrative requirements of counting, not vice versa.

4.6. Wrap-up

We human beings are "florid" representational creatures. We use language, thought, and writing in interleaving ways to represent ourselves and the world around us. For instance, we write down what we spoke, but we also think aloud, we narrate, read, quote, and memorize texts, and we use word-processors to compose philosophical arguments. Much of advanced human problem-solving depends on the skillful coordination of internal and external representational resources. Learning how to become a "florid" representational creature who can productively and effortlessly switch back and forth, in a seemingly open-ended spiral, between different representational systems may be one of the greatest achievements of language-infected biological brains like ours. When we thus shuffle, translate, and constructively recombine symbolic patterns from one medium to another, the cognitive synergies arising from our representational activities can outstrip the information that is contained in any of its parts. Because of these often spectacularly transformative effects, I conclude that the *representational re-description of information in different formats is (in many cases) itself a cognitively significant constructive process*.

My goal in this chapter was to argue against the "assumption of cognitive neutrality" as it is manifested in three related doctrines about mind, language, and cognition:

1. Artifacts are merely external aids for thinking.
2. Language is merely a vehicle for expressing thoughts.
3. Writing is merely a transcription of speech.

Against (1), I have argued that much of advanced human cognition is not achieved by the "naked" brain alone, but intimately dependent on the active contributions of a wide array of culturally constructed artifacts that are reliably available, developmentally basic, and easily accessible in the modern environments that we have designed and in which we grow, think, and act. Against (2), I have argued that both spoken and written language are powerful mind-tools. By way of contrasting the views of Dennett and Clark, I have distinguished two ways in which our acquaintance with external lingua-form resources has reshaped our cognitive capacities. According to Dennett's more developmental account, acquiring a language significantly *alters* the internal computational architecture of the brain, by installing a new kind of "virtual machine." According to Clark's emphasis on the hybrid nature of extended cognition, language mostly works its magic as a cognitive artifact that *complements* (but without deeply altering) the basic representational and computational resources of our biological brains. Both views concur that the cognitive involvement of language supports the development of otherwise unavailable forms of expertise, and underpins our meta-cognitive skills. Finally, I have suggested that there is a residual "written language bias" in Clark's conception of language as a cognitive artifact, and that some of the cognitive benefits discussed by Clark really arise only from literate practices. Expanding on this idea, I have argued against (3) that writing systems (in particular, the alphabetic script) are *sui generis* types of cognitive artifacts that do not merely recapitulate the representational roles of spoken language, but provide new forms of data processing and manipulation that transform the profile of extended cognitive agents.

5. Extended reasoning

5.1. A representational argument for the HEC

Edwin Hutchins (1995a, p. 114) once described the importance of understanding the nature of tool use as follows:

> "Every argument showing why a particular tool is easy to use is also an argument showing why both internal and external tools are part of the very same cognitive ecology. *It is a truism that we cannot know what the task is until we know what the tools are*. Not only is this true of both internal and external tools, it is also true of the relationships among them" (my emphasis).

In this chapter, I shall argue that Hutchins' astute analysis – a "truism," as he calls it – can be turned into an argument for the hypothesis of extended cognition (HEC). To sneak up on what I will call the *representational argument* for the HEC, let us first distinguish several different levels on which to interpret Hutchins' insightful remarks.

On the most obvious level, the main point of Hutchins' analysis is to promote the design of good user-interfaces for cognitive artifacts. It rests on the by-now familiar consideration that by mediating between a task and a user, a tool transforms the nature of the work which the user has to perform (see Chapters 2.5, 4.2 for discussion). As an example of the suggested transformation, Hutchins discusses the widespread use of nautical slide rules for the purpose of ship navigation (1995a, Chapter 3). A slide rule is an extremely useful device for people who, like navigators, are frequently in a pinch trying to solve cognitively taxing distance-rate-time problems (e.g., computing the speed of a ship based on the distance it has travelled during a fixed time interval). Courtesy of the mechanism that is built into the structure of the slide rule, but effectively hidden from the user, the operations by which a person aligns the scales of her slide-rule lie in the perceptual-motor domain; however, the computations accomplished by the combined person-artifact system lie in the domain of arithmetical reasoning. What makes the slide rule a well-designed piece of cognitive technology is that our biological brains are much more apt at perceiving and manipulating physical objects in space than they are at manipulating arbitrary strings of symbols – regardless of whether those symbolic manipulations are done on paper or in one's head (Norman, 1991).

On a deeper level, Hutchins also intends his remarks as a methodological critique of the orthodox, symbol-based computational approach in cognitive science. According to Hutchins, the fundamental error of symbol-based cognitive science has been to mistake the properties of socio-culturally mediated person+artifact systems for a model of the human brain (Hutchins 1995a, Chapter

9). For instance, suppose that as cognitive psychologists, we neglect or abstract away from the fact that in their concrete work environment *as* navigators, human beings usually employ a slide rule to solve navigational problems. Instead, we apply the same computational analysis of navigational problem-solving to the "naked" human brain, as if it were solely responsible for calculating the ship's speed. In doing so, we'd subtract the work done by the slide rule from the computational analysis of an entire person+artifact system, but without revising our description of the task done by the person. This is bound to yield a systematically distorted view of the operations which the user typically performs, as a navigator, in her concrete task environment. In particular, a cognitive psychologist working in her lab might be tempted to attribute to the *brain* a series of symbolic computations that in reality are distributed between brain, body, and external symbolic artifacts. Psychologists have a nice term to characterize situations in which the setting of a study fails to approximate the real-life situation it is meant to investigate: they call it a lack of "ecological validity." According to Hutchins, orthodox cognitive-science suffers from a severe lack of ecological validity in its tendency to commit "cognitive overattribution" fallacies, i.e., to "put much more inside [the head] than there should be" (ibid., p. 356).

Hutchins' methodological critique seems particularly relevant and timely when we look at the cognitive psychology of deductive, especially syllogistic reasoning. Because of its brain-bound methodology (see Chapter 6.2 for further discussion), the mainstream psychology of deductive reasoning has for the longest time downplayed (if not altogether ignored) the impact of culturally constructed representation systems on our formal reasoning abilities (cf. Stenning, 2002; Stenning & van Lambalgen, 2008). This may well be a legacy of the received "rationalist" view that our ability to reason is a fixed part of our cognitive endowment as rational animals. For instance, it has often been assumed in the psychology of syllogistic reasoning, typically without further argument, that people are essentially homogeneous in what kinds of mental representations they use when they reason; or that visually appealing features of external notations have no direct bearing on our "amodal" system of mental representations; or that learning how to reason is essentially a matter of tapping into some universal, biologically predisposed inferential mechanism. Evidence for the lasting influence of such assumptions can be spotted in a number of otherwise competing theories of syllogistic reasoning – such as the theories of Mental Models (Johnson-Laird, 1983), Mental Logics (Rips, 1994), Euler Circles (Stenning & Oberlander, 1995), the Language of Thought (Fodor, 1975), or the innate modules of evolutionary psychology (Cosmides, 1989). Surprisingly little attention has been paid by psychologists to how students *learn* to reason in highly "scaffolded" environments such as our classrooms, and why students systematically differ as to what kinds of representational systems (e.g., sentential vs. diagrammatic systems) they prefer to use.

In *Seeing Reason*, Keith Stenning (2002) has spearheaded an alternative "constructivist" approach to reasoning. It acknowledges the critical role of selecting an appropriate external representation system, and also the constructive re-organization that occurs within a learner when she acquires the skills to use a new formalism. Summarizing his findings, Stenning concludes that "[...] there is more than one mental representational approach to syllogisms" (ibid., p.128), suggesting that the traditional assumption of representational homogeneity is no more warranted in psychology than it is in logic, where we have a large variety of different systems for representing syllogistic inferences.

Stenning's constructivist approach raises an interesting meta-theoretical problem which finally leads us to the third and potentially most consequential level on which we can interpret Hutchins' remarks about tool use. At the very least, Stenning's research program calls for more systematic investigations of the cognitive effects which different types of external representation systems (e.g., sentential vs. diagrammatic systems) have on our reasoning abilities. As he points out in a related context (ibid., pp. 17f), it is clear that such comparative investigations require that there is *some* relation of invariance under which the compared processes of reasoning are cognitively equivalent, in the sense that they all pertain to a given domain of inferential processes (e.g., the syllogistic domain). But what facts justify the claim that the observed processes are all instances of (e.g.) *syllogistic reasoning*? If the traditional assumption of representational homogeneity is correct, there is an easy answer to this question. We can simply point to the hypothesized class of privileged mental representations (e.g., Mental Models, Mental Logics, or Euler Circles) in the head that people invariably use whenever they are engaged in syllogistic reasoning. But if Stenning is right, there is no such universally shared representational mechanism in the brain for syllogistic problem-solving. So what other facts – if any – about our cognitive machinery make all these processes instances of *syllogistic reasoning*? More generally, *what grounds the domain-identity of cognitive processes when the neural mechanisms by which those processes are performed are evidently not domain-identical*?[53]

53 In Chapter 2.3, I made a similar point about the interpretation of functional similarity in the context of Clark and Chalmers' (1998) parity principle.

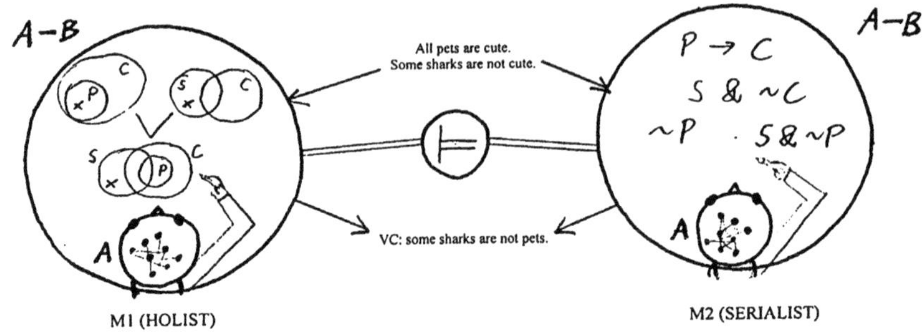

Figure 15. An illustration of the meta-theoretical problem of cognitive domain-identity. Left: A HOLIST is using EC (Euler Circles) to solve a syllogistic problem. Right: A SERIALIST is using a sentential fragment of ND (Natural Deduction) to solve a syllogistic problem. In each case, the system labeled "A" represents the "narrow" part of the computational mechanism that is located in the brain; the larger system labeled "A-B" represents the "wide" computational mechanism that stretches beyond the brain (and body) to include parts of the environment.

A standardized presentation of the suggested meta-theoretical problem will help to facilitate our subsequent discussion (Figure 15). Suppose that two people (let's call them "HOLIST" and "SERIALIST") belong to different populations of learners who prefer to reason with different types of external representation systems, perhaps because of their different background skills and cognitive styles.[54] To solve syllogistic problems, HOLIST thus learns to use Euler Circles (EC) which suit her reasoning style, whereas SERIALIST learns to use a system of Natural Deduction (ND) which suits her distinct style.[55] As representational systems, EC and ND each belong to a different representational modality. EC is a diagrammatic method of reasoning, whereas ND is a sentential method of reasoning (Stenning, 2002). Both eventually become expert problem-solvers in their respective domains. When asked to solve syllogisms using pen and paper, HOLIST deploys a modality-specific computational mechanism M1 which operates on external EC, utilizing a class IR1 of internal representations. SERIALIST, on the other hand, deploys a different modality-specific computational mechanism M2 which operates on external ND, utilizing a distinct class IR2 of in-

---

54 See Pask (1975). Monaghan and Stenning (1998) report strikingly symmetrical aptitude-by-treatment interactions between HOLISTS and SERIALISTS in their assessment of teaching/learning-effectiveness with diagrammatic versus sentential methods; see also Monaghan et al. (1999).

55 Stenning and Yule (1997) present a case-based algorithm for syllogistic reasoning which uses only the sentential fragment of the natural deduction calculus. See below for further discussion.

ternal representations. M1 is designed to process the spatial, geometrically structured features of EC, but cannot process information presented in ND. In contrast, M2 is designed to process the symbolic, syntactically structured features of ND, but cannot process information presented in EC. What makes our supposition that SERIALIST and HOLIST both engage in *syllogistic reasoning* true? Providing a positive solution to the meta-theoretical problem of cognitive domain-identity suggests the following representational argument for the HEC (cf. Theiner, 2007):

1. SERIALIST and HOLIST both engage in syllogistic reasoning.
2. The claim that SERIALIST and HOLIST both engage in cognitive processes of syllogistic reasoning is made true by the fact that they manipulate mental representations which have the same syllogistic content.
3. In the context of the task, the only relevantly available representations are {IR1, EC} for HOLIST, and {IR2, ND} for SERIALIST.
4. Tokens of IR1 and IR2 are semantically type-distinct when considered in isolation.

Therefore, the only mental representations which have the same syllogistic content are external tokens of EC and ND.

Therefore, the HEC is true.

The truth of the HEC follows from the the fact that the external representations on paper are literally parts of extended reasoning processes. For the purpose of discussion, I shall take our supposition in Premise (1) at face value. In the next three sections, I will defend the remaining three premises of our argument in the following order: first (2), then (4), and finally (3).

5.2. On the domain-identity of cognitive processes

Premise (2) calls upon the content of mental representations to ground the domain-identity of cognitive processes. It is based on the principle that the domain D of a cognitive process, carried out by some psychological mechanism M, is determined by the informational content of the mental representations on which M operates. By this criterion, token-processes of the type *face recognition* are thusly individuated because they are carried out by a psychological mechanism which processes representations of the configurational properties of faces. Similarly, Premise (2) asserts that *syllogistic reasoning* is individuated in terms of representational mechanisms that are capable of tracking the relation of logical consequence for a well-defined fragment of monadic first-order logic.

A full-blown defense of the computational-representational theory of mind (Horst, 2011), on which Premise (2) rests, is clearly beyond the scope of this chapter. I will thus proceed indirectly, by considering several versions of an alternative criterion that is based purely on input-output equivalence. I shall argue that the alternative criteria all succumb to a dilemma. In their strictly behavioral versions, they do not solve our meta-theoretical predicament about cognitive domain-identity. But in suitably constrained representational versions, they actually support the HEC.

In order to provide a real alternative to our representational criterion, we first need to establish a manner of speaking about cognitive states and processes in a way which does not ontologically commit us to the existence of mental representations. If we consider mental representations as theoretical constructs of cognitive science, one way this could be done is by employing the Ramsey-Lewis method for defining theoretical terms (Lewis, 1970, 1972). Using the Ramsey-Lewis method, we use a meta-language in which the vocabulary for denoting external stimuli and behavioral responses is treated as observationally primitive, and mental states are interdefined functionally as states that causally interact with each other and sensory stimuli to produce behavioral responses. Based on this method, let us consider a purely behavioral (i.e., non-representational) criterion of domain-identity. It states that two cognitive processes – understood as causally ordered state transitions between cognitive states – are domain-identical just in case they both map the same class of sensory inputs onto the same class of behavioral outputs. Notice that this would be a behavior*al*, but not a behavior*ist* criterion of cognitive domain-identity. Applied to our example, it does not deny that there are occupants of the functionally individuated mental states previously described as "IR" and "ER," and that those states causally mediate the mapping from inputs to outputs. However, these internal states are considered to be irrelevant for the semantic specification of the domain to which a given cognitive process pertains.

To see whether the suggested behavioral criterion can be used to solve our meta-theoretical predicament, let us try out more specific proposals of what the relevant "inputs" and "outputs" of the mechanisms M1 and M2 might be.

Proposal 1

According to our first proposal, the relevant input is the entire class of syllogisms (expressed in a given language, e.g., in English), and the relevant output is the response consisting of either 'NVC',[56] or 'VC' plus the available conclusion(s), expressed in the same language. Wedged in between input and output is the psychological mechanism which is causally responsible for computing the desired mappings from inputs to outputs. A subset of the states which are part of

56 'VC' stands for 'valid conclusion,' 'NVC' for 'no valid conclusion.'

such a mechanism match the functional profile of mental representations, as described with the help of the Ramsey-Lewis method. Since there are no further constraints in place on what counts as a mental representation, both IR and ER are in principle equally capable of satisfying the relevant functional roles. Since we assume that M1 and M2 are in fact input-output-equivalent, the principle of parity (Chapter 2) implies that it should not matter whether the relevant functional roles are all satisfied internally, or whether they are distributed across internal and external resources. From Premises (3) and (4), we know that members of the class of internal representations (IR) are not sufficient as occupants of all the required functional roles. Hence the syllogistically relevant mental representations must, in M1 and M2, be distributed across IR and ER. This recapitulates, of course, our considerations in favor of *extended functionalism* (Chapter 2).

In sum, Proposal 1 correctly implies, consistent with Premise (1), that HOLIST and SERIALIST are both reasoning syllogistically. But it would also validate the HEC in virtue of the distribution of functionally relevant causal labor between inner and outer resources in the extended mechanisms M1 and M2. I conclude that a purely behavioral criterion of cognitive domain-identity can hardly be used to resist our representational argument for the HEC.

Proposal 2

A second proposal about the relevant "inputs" and "outputs" for identifying the type of a cognitive domain is to break down the global input-output mapping into a series of simpler mappings. For instance, it seems natural to break down the operation of M1 and M2 into three parts (or stages). During the first stage, syllogistic arguments have to be translated into one of the two specialized notations. During the second stage, the external representations have to be translated into some internal representational format used by the brain, processed, and then written out on paper. During the third stage, the outputs of the second stage are translated backwards into 'NVC,' or into 'VC' plus the relevant conclusion(s) in English. For the sake of concreteness, let us suppose that the suggested decomposition of M1 and M2 looks as depicted in Table 2.

Table 2. Functional decomposition of the input-output mapping computed by M1 and M2.

	M1 (HOLIST)	**M2 (SERIALIST)**
Input 1	Class of syllogisms	Class of syllogisms
Output 1 = Input 2	Two premise diagrams in EC	Two premise translations in ND
Output 2 = Input 3	Agglomerated diagram in EC	Completed derivation in ND
Output 3	'NVC,' or 'VC' + conclusion	'NVC,' or 'VC' + conclusion

In our computational practice, the second stage of the computation is likely to involve a series of highly interactive cycles of perception and action in which notational inputs are repeatedly "caught" by the brain, internally manipulated, and subsequently "tossed" back to paper.[57] Despite the densely interactive causal traffic between brain and world, critics of the HEC have argued that in principle, we can always analytically separate the brain-bound cognitive resources from the environmental *non*-cognitive resources without any loss of descriptive or explanatory power (Adams & Aizawa, 2008; Rupert, 2009; cf. Chapter 1.4). According to this deflationary interpretation, cognition would be strictly brain-bound, even though the environment functions as a temporary storage for offloading information.

However, the suggested functional decomposition of M1 and M2 does not solve our meta-theoretical predicament. This is because the hypothesized sub-processes performed by the brain (during the second stage) are not strictly input-output equivalent in the required sense. By hypothesis, the SERIALIST's M2 is not able to process the graphical representations of EC. This means, for instance, that if we took Input 2, as represented in EC, and would feed it into the HOLISTS's M2, M2 would not be able to compute the desired Output 2. Likewise, M1 could not compute any of the required functions if its inputs are symbolically expressed in the form of ND. Since the hypothesized sub-processes are evidently *not* input-output equivalent in this case, it would follow that the cognitive processes in which HOLIST and SERIALIST are engaged do *not* belong to the same cognitive domain. This contradicts Premise (1).

Proposal 3

To purge the behavioral criterion from its strong dependence on specific representational features of a given notation, the critic will argue that it ought to be applied to "syllogistically relevant" equivalence classes of inputs and outputs. One way this might be accomplished is by defining equivalence classes between specific types of inferential operations afforded by EC and ND. For instance, the case-based method of ND presented by Stenning and Yule (1997) is designed to "emulate" the ability of EC to track *critical individuals*.[58] Case-based reasoning methods work well for the syllogistic fragment of first-order logic, because every syllogistically valid inference is captured by a unique model that is defined by a single critical individual.[59] This property is known as the "case-

57 The "catch-and-toss" metaphor is borrowed from Clark (1997, p. 105).

58 In a syllogistic inference, critical individuals are things which are fully determinate with regard to their membership in the classes expressed by all three terms.

59 Because of this meta-logical property, the truncated variant of ND introduced in Stenning and Yule (1997) does not need any quantifiers, and has only modus ponens and modus tollens as its rules of inference.

identifiability" of syllogistic logic (Stenning & Oaksford, 1993; see below for further discussion). For present purposes, Stenning and Yule's analysis could be used to define broader equivalence classes of inputs and outputs that are independent of being expressed in a specific representational modality. As a result, one might say that the brain-bound portions of M1 and M2 are domain-identical, because they can be used to compute the same equivalence class of input-output mappings.

However, this proposal for establishing the cognitive domain-identity of M1 and M2 is evidently parasitic on the existence of strong representational homologies between EC and ND which have been put there by design. They derive from their shared *semantic* ancestry as case-based methods. The implicit dependency of their design on semantic considerations has two important consequences.

First, the third proposal is no longer a purely behavioral criterion of cognitive domain-identity, but is really at bottom a semantic and thus *representational* criterion. This can be shown by applying David Marr's (1982) three-level computationalist framework for analyzing cognitive systems to M1 and M2. According to Marr's framework, the main goal of a Level-1 analysis of complex information-processing systems is to pin down their behavior in terms of a precise input-output function. But an algorithmic Level-2 analysis goes beyond mere input-output functions. A Level-2 algorithm must provide explicit schemes for *representing* syllogistic inputs and outputs, together with a sequence of mechanical steps to *execute* the relevant subtasks. Based on Marr's distinction, a Stenning-and-Yule-type reconstruction of EC and ND amounts to a full Level-2 analysis of M1 and M2.

This leads us to the second implication. Once it is conceded that we have to descend to Level-2 analyses of M1 and M2 to account for their semantically defined domain-identity, proposal 3 collapses into a form of *extended computationalism* (Wilson, 1994), which itself is a specific version of extended functionalism and thus supports the HEC. Once again, this follows from the principle of parity (Chapter 2). Within Marr's framework, the representational structures to which we refer in Level-2 explanations of information-processing tasks play the theoretical role of mental representations when the analysis is applied to the brain. Because of parity, our explanatory practice should not change if we move our relevant unit of computational analysis beyond the head into the world on paper. If we appeal to mental representations to ground the identity of a cognitive domain, we should grant that external representations are equally apt to play that meta-theoretical role. In fact, external representations may be the *only* vehicles capable of playing that role if we can further show – as claimed in Premise (4) – that the *internal* representations of HOLIST and SERIALIST are ill-equipped to ground the domain-identity of our two token-processes of syllogistic reasoning. Hence defending Premise (4) is the next task to which I shall now turn.

5.3. A "division of cognitive labor" hypothesis

When mental representations are posited as theoretical entities in psychological explanations of behavior, they are expected to do double-duty (Fodor, 1985): first, they ought to be semantically evaluable, i.e. to serve as stand-ins which are *about,* or *refer to*, certain things; second, they ought to play a causal role in the production of intelligent behavior. These two requirements constrain the kinds of states which are apt to occupy the role of mental representations. Considering *aboutness* as a distinctive mark of the mental, Adams & Aizawa (2001, 2008) have argued against the HEC that environmental resources are never cognitive because their intentional content is always merely derived from the content of mental representations inside one's head (see Chapter 2.10). Challenging this claim, I shall now argue that in the the scenario that I have described, neither internal nor external resources by themselves have the right Janus-faced profile to satisfy both the semantic and the causal requirements at once. As a result, only the functional integration of their joint contributions qualifies the extended computational process as an episode of *syllogistic reasoning* in good standing. For the sake of brevity, I will focus on the role of Euler Circles (EC) in the HOLIST's mechanism M1 to expound my "division of cognitive labor" hypothesis.

To begin with, let us briefly consider some central features of EC as a graphical algorithm for syllogistic reasoning. Solving syllogistic problems in one's head is hard, because it places a high cognitive load on our working memory. It requires the temporary binding of properties into specifications of individuals, grouping these individuals into models, and often comparing several alternative models. These bindings have to be constructed from scratch for each problem, without being able to rely on previously established bindings that are stored in long-term memory. In light of these demands, the direct semantic interpretability of spatial relations in EC (Figure 16) significantly facilitates the processibility of syllogistic inferences for our brains in a number of ways (based on Stenning & Oberlander, 1995; Stenning, 2002).

First, taking the spatial containment of a point in a curve to denote set-membership, the topology of EC enforces a fully determinate representation of an individual's properties, since every point in a plane is either inside or outside a closed curve. The EC-notation thus provides a perspicuous representation of the required bindings that is easy to attend to. Assuming that all closed curves are continuous and refer uniquely to properties, EC are *self-consistent* insofar as no single diagram can represent an inconsistent set of propositions.

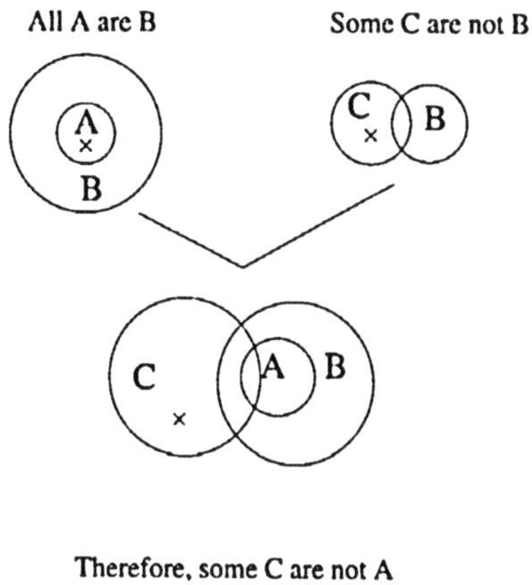

Figure 16 (from Stenning, 2002, p. 99). Representing a valid syllogistic argument with the help of Euler Circles

Second, to distinguish individuals that *must* exist (in a universe of discourse) from those that only *may* exist, HOLIST uses the cross-notation as an "abstraction trick" to represent only minimal models of each premise. To combine two premise diagrams, HOLIST registers the B-circles, choosing the arrangement between A- and C-circles which creates the most subregions. If a cross-marked region is bisected during this agglomeration, the cross is excised. A persisting cross in the final diagram thus indicates a valid inference.

Third, by allowing the brain to treat circles as "hoops" and crosses as "nails," EC act as a quasi-mechanical device that mirror the logical constraints imposed by the premises. For instance, if a "nail" prevents the A- and C-circles from being pulled apart, at least one positive conclusion can be drawn; if a "nail" prevents a complete alignment, at least one negative conclusion can be inferred. HOLIST can thus also bring her prior mechanical knowledge to bear on syllogistic tasks. Moreover, a simple graphical heuristic allows HOLIST to determine whether the premises entail a universal or a particular conclusion. A universal conclusion requires that the critical region be circular and labeled by an end term, which turns into the subject of the conclusion. Otherwise, if the argument has already proven to be valid, at least one particular conclusion can be formed by dropping the B-term.

Fourth, comparing the meta-logical properties of EC and ND, we can show that they are both sound and complete as syllogistic proof procedures, and that there are close representational homologies between the modality-specific operations afforded by the two formalisms (Stenning & Yule, 1997).

In sum, EC are well-designed cognitive artifacts because their geometrical and pseudo-mechanical properties effectively mirror the logical constraints of the syllogistic domain. They act as external resources which allow HOLIST's brain to accomplish syllogistic computations by performing operations which lie in the visuo-spatial domain.

On the other hand, I shall now argue that there is nothing intrinsically *syllogistic* about the neural processes going on inside HOLIST's brain when she performs the suggested visuo-spatial manipulations. In order to bring out the semantic contrast between inner and outer representations, let us consult some leading "naturalized" theories of intentional content – first, as they apply to the external representations of EC, then as they apply to the internal representations that are hypothesized to reside inside the brain. Two established naturalistic theories of mental content seem to be particularly well-suited to allow such a comparison: the theory of indicator semantics developed by Fred Dretske (1986, 1991), and the teleo-semantic account of mental content developed by Ruth Millikan (Millikan 1984, 1989).

As a concrete working example, let us consider the claim that the cross-marked region in the EC-diagram of a conclusion is an externally located mental representation of the *unique minimal model* (UMM) which always captures a valid syllogism. According to Dretske's teleo-informational theory of mental content, a mental state R represents P just in case R has an acquired function for the behavior of a cognitive system, and it acquired that function *because* it indicated P. Dretske's notion of indication is cashed out in terms of conditional probabilities: a signal A *indicates* or carries information about a state of affairs B just in case Pr (B|A) = 1.[60] Dretske requires that R must have been specifically recruited by a cognitive system for the causal ("structuring") role which its indicational properties play in the production of behavior. Dretske's teleo-informational account of mental content can be straightforwardly applied to the semantics of EC. It correctly implies that M represents UMM because it has an acquired causal role in the process by which a person solves syllogistic problems, and it acquired that function only because it reliably indicated the presence of UMM. For instance, the presence of M leads HOLIST to utter a "valid conclusion" response, whereas its absence leads her to say that there is no valid conclusion. This response pattern is plainly an acquired psychological function, since it is unavailable to a person who has never learned how to reason with EC.

A different class of teleo-semantic theories invokes a teleological notion of *function* to determine the semantic content of a mental representation (Neander,

60 Dretske's strict requirement that Pr (B|A) = 1 in all cases leads to some complications in his theory that we can safely ignore here.

2009). On a general level, teleo-semantics is the claim that a structure R represents a property P just in case R's bearing some privileged relation to P accounts for the continued reproduction of R. In order to describe the nature of this privileged relation, Millikan (1984, 1989) has appealed to the functional contribution that a representation R makes to the system consuming R in order to perform its *proper function*. A consumer system has a proper function if tokenings of that function do something that has positively contributed to the survival and proliferation of that system in a population. The point of invoking proper functions is to establish *Normal* success conditions which are apt to determine the non-derived representational content of R (Fodor, 1990). For instance, consider the much-cited case of the frog that responds to the presence of a fly in its visual field by lashing out its tongue (Lettvin et al., 1959). According to Millikan's theory, the frog's visual representation R represents *frog food*, and not *anything appropriately small, dark and moving past its retina*, since it was only when frog-food triggered tokenings of R that the frog was fed, and only then did the frog's digestive system contribute to the frog's survival and reproduction by using R.

Although Millikan's theory of proper functions has originally been formulated within the biological context of natural selection, the founding insight of teleo-semantics can be preserved if we invoke a process of *cultural* selection instead to establish appropriate Normal conditions. In the past two decades, there has been considerable success in applying the Darwinian logic of evolution-by-selective-retention to explain cultural variation (Richerson & Boyd, 2005). It seems natural to think that a process of cultural selection would account for the selective adoption of EC by our population of HOLISTS. With this relatively minor modification, we can apply teleo-semantic theories to culturally evolved representation systems. In particular, we can say that M represents UMM since EC have been culturally selected by HOLISTS as their preferred syllogistic representation-producing device, and it was only because M corresponded to UMM that the consumption of EC has contributed to the adaptive success of HOLISTS as syllogistic reasoners.

Now consider how the same theories of mental content imply a different outcome when we apply them to the hypothesized class of inner mental representations. According to the constructivist approach that we have outlined in the beginning, how should we think about the nature of those mental representations? Commenting on the constitutive role of external representation systems, Stenning (2002) notes that "at the very least one would expect external representations to obviate the need for some internal memory representations, replacing them by perceptual-motor processes with their own ephemeral representations" (p. 124). Following the footsteps of Barsalou (1999), let us consider these "ephemeral" representations to be perceptually grounded working memory representations that reflect the brain's previous sensory and proprioceptive experience of interacting with external objects – including three-dimensional things like disks, circles, but also dealing with graphical representations (see also Chapter 4.5.2). So let us make it part of the description of our thought experiment that the mental representations which HOLIST uses to manipulate EC are perceptually fine-tuned to various visually distinctive features of the external notation. Considered in isolation, these representations need not have any intrinsic syllogistic significance at all. Instead, their proper cognitive function is to depict external representational structures in terms of how to manipulate them effectively "with eyes and hands" (Latour, 1986).

For instance, the precise shape, distance and positional placement of an A-circle vis-à-vis some B-circle *on paper* is logically insignificant as long as (say) A is properly contained in B. However, the *neural* representation of their exact spatial configuration might determine whether it is easier for a subject to internally "move" the first premise diagram on the second diagram when she compounds the two, or rather do it in the opposite direction. Let us assume that in the course of reasoning, the mental representations in HOLIST's brain depict the EC-notation primarily in terms of "PIN-PREVENTS-SMALLER-RIGHT-HOOP-FROM-BEING-HORIZONTALLY-PULLED-APART-FROM-UNDERLYING-BIGGER-LEFT-HOOP." On the other hand, the mental representations in SERIALIST's brain depict the ND-notation primarily in terms of "SHAPE-OF-SMALLISH-*B*-SPACED-PRETTY-CLOSE-TO-'→'-MATCHES-SHAPE-OF-MUCH-BIGGER-*B*-IN-LINE-BELOW." Thanks to the perceptually grounded nature of those mental representations, our subjects co-opt their preexisting background skills for the purpose of learning how to manipulate a new representational formalism. But from a semantic perspective, it reveals our old meta-theoretical predicament. If we abide by our representational criterion of individuating cognitive processes, the within-the-head parts of M1 and M2 constitute two distinct perceptual-motor processes, and not two tokens of *syllogistic reasoning*.

We can further underscore the semantic asymmetry between inner and outer representations in this context by employing the notion of a *proprietary domain*. The proprietary domain D of a cognitive mechanism M is the type and range of information which M has been designed to process,[61] because it involves the manipulation of mental representations whose proper function is to carry information about properties of D (cf. Millikan, 1984). Using this notion, we can de-

61 I shall leave it open here whether the design of the cognitive mechanism is a result of biological evolution, individual learning, or cultural evolution.

fine the proprietary domain of a cognitive mechanism in terms of the *intrinsic* semantic content of the mental representations on which it operates.

According to this definition, the within-the-head part of M1 is clearly not proprietary to the syllogistic domain. We acquire our cognitive abilities to understand overlaps, detect bisections, and recognize constraints on the movement of physical objects by doing everyday tasks in a normal environment. It does not require that we ever set foot in a classroom where we are first exposed to EC, and we can assume that the same is true of HOLISTs. It follows that the syllogistic content of internal representations (IR) during pen-and-paper manipulations is only *derived* from the syllogistic content of external representations (ER). To see this, consider, once again, the same theories of mental content to which we have appealed before. In Dretske's terms, IR has indirectly acquired the cognitive role within M1 to detect UMM *only* because ER reliably indicates the presence of UMM. In teleo-semantic terms, IR has indirectly acquired the function of representing UMM within M1 *only* because ER corresponds to UMM, which was the reason why EC have originally been culturally selected by HOLISTS as their preferred syllogistic representation-producing device.

In conclusion, let me succinctly summarize the implication of my "division of cognitive labor" hypothesis between internal and external representations. External representations, it is safe to say, are dependent on perceptual-motor and neural mechanisms to guide our behavior. But the perceptually grounded mental representations which people use to manipulate external inferential mechanisms are, in turn, dependent on external representations to acquire genuine syllogistic (as opposed to, say, visuo-spatial content) content. It is only because we consider the within-the-head part of M1 *together* with its beyond-the-head part as a single, hybrid system, that M1 can properly be viewed as an extended mechanism of *syllogistic reasoning* in good standing; the same sort of considerations apply to M2. According to the representational criterion of cognitive domain-identity suggested in Premise (2), our claim that HOLIST and SERIALIST both engage in episodes of syllogistic reasoning is therefore semantically accounted for by the fact that they manipulate type-identical *mental representations* (EC and ND) that are located outside their heads. Therefore, the syllogistic reasoning of HOLIST and SERIALIST stretches beyond the head into the world.

5.4. The role of meta-representational knowledge

Let us now turn to Premise (3) which – at first blush – appears to be the least controversial premise of our argument. Given how we set the stage for the suggested comparison between HOLIST and SERIALIST, what representations other than {IR1, EC} and {IR2, ND} could possibly ground the domain-identity of the two extended computational processes? But first appearances can be deceiving. Our argument rests on the claim that IR1 and IE2 are embodied, perceptually grounded, "ephemeral" mental representations that are recruited by the brain for syllogistic reasoning tasks. While there is considerable evidence that a significant portion of mental representations are like that, it is less clear that *all* internal representations are like that. In particular, it remains open to the psychological internalist to argue that my assumption fails to consider the rich meta-representational knowledge which competent human reasoners must also bring to the task.

This potential lacuna can be turned into the following objection to the presently explored representational argument for HEC.[62] When we teach students how to reason with EC or ND, we not only instruct them how to manipulate a bunch of meaningless squiggles, but also aim to impart a deeper understanding of the inferential roles that are played by circles, crosses, overlaps, shaded regions, and so on in those algorithms. Students who are capable of understanding such explanations must therefore be able to entertain fairly rich meta-representational beliefs about the syllogistic significance of various aspects of the notation which they are using. As a case in point, we typically regard students who can flawlessly execute a graphical or a sentential algorithm without showing any evidence of such deeper understanding as cognitively deficient reasoners. Provided that we consider SERIALIST and HOLIST as fully competent reasoners, it is the semantic content of their meta-logical and meta-representational beliefs which ultimately grounds the syllogistic content of the external formalisms. Since those meta-representations are realized inside a person's head, this shifts the criterion by which we individuate the domain of syllogistic reasoning back in favor of psychological internalism. To put it bluntly, and echoing Hutchins, we would commit an "underattribution" fallacy if we decided to neglect the rich meta-representational knowledge possessed by competent syllogistic reasoners.

I share the concern that is part of the sentiment behind this objection. It brings out an important point that is sometimes overlooked in debates over the extended mind. Proponents and critics of the HEC can agree that the strategic manipulation of external tokens to solve problems of deductive reasoning is a paradigmatic case of *symbolic off-loading* (Chapter 1.2.2). More importantly, they can agree that symbolic off-loading is an instance of a "florid" (Dennett, 2000) representational activity, i.e., a deliberate, knowing, self-conscious cognitive practice which is a distinctive strength of our species (see Chapter 4.4). As Dennett has plausibly argued – and his point is reflected in the above objection – there can be no florid representation without meta-representation. Similarly, it

62 A response along these lines has been suggested to me by Robert Rupert (pers. comm.).

would be foolhardy to deny that expert reasoners such as HOLIST and SERIALIST have the meta-representational capacities to interpret external tokens *as* standing for certain premises or inferences in a syllogism. For example, Stenning (2002) has argued that a learner's pre-existing forms of meta-representational knowledge (e.g., about diagrams, alphabetic writing systems) determine the starting point of her learning curve, and this may account for inter-individual differences in students' preferences for either diagrammatic or sentential formalisms (Stenning, 2002, p. 131). Hence it is important to rectify a common misperception about the HEC. When HEC-theorists highlight the cognitive benefits of symbolic off-loading, they do not mean to deny that people also have to acquire a proper understanding of how to interpret those external representational structures. On the contrary, the internalization of cognitive norms is an indispensable prerequisite for the successful cognitive integration of internal and external resources (Menary, 2007; Sutton, 2010).

It is therefore fully compatible with our argument that a centerpiece of the constructivist approach advocated by Stenning is that "theories of human reasoning must incorporate accounts of metalogical knowledge" (2002, p. 124). A familiarity with specifically designed notational systems can, under the right circumstances, trigger new meta-logical insights in the learner. For instance, case-based graphical methods like EC may be especially well-suited (at least for HOLISTs) to reveal an important meta-logical property of syllogistic logic – its *case-identifiability* – because they exploit a learner's more basic geometrical intuitions about diagrams (Stenning, 2002, pp. 128-34).

A striking demonstration of the sensitivity of syllogistic problem-solving to meta-logical knowledge comes from a study by Stenning and Yule (1997). There is a special class of syllogisms which yields so-called "*u*-conclusions"[63] (ibid.) that would produce a cross in EC, but there is no proper way of drawing these conclusions in the conventional Aristotelian framework.[64] This raises the question of whether naïve subjects have an intuitive grasp of the model that captures these conclusions.[65] Stenning and Yule were able to show that depending on the way in which a syllogistic production task is framed, subjects can be

---

63 The u-conclusions, such as "Some not-A are not C," involve a fifth type of quantifier (abbreviated 'u') in addition to the four a-, e-, i-, o-type quantifiers recognized by Aristotle (Stenning & Yule, 1997). For instance, from No A are B, Some B are not C, we can infer that "Some not-A are not C."

64 This shows that the converse of case-identifiablity – i.e., whenever there is a cross, there is a valid inference – fails to hold in the Aristotelean framework.

65 This experiment was performed with naïve subjects who were not previously taught how to use external representation systems to solve syllogistic problems.

primed to respond either way. When they were instructed to perform an *individual description task* (i.e., to determine if the syllogism implies the existence of critical individuals), about half of the subjects correctly produced the *u*-conclusion (i.e., as VC). On the other hand, when the problem was framed as a conventional task of drawing a conclusion which relates A and C by using one of the four Aristotelian *a*-, *e*-, *i*-, *o*-type quantifiers, about two-thirds said that there is no valid conclusion that can be drawn. These results strongly suggest that the subjects' first-order reasoning abilities were affected by higher-order meta-logical beliefs, which in turn presupposes fairly sophisticated meta-representational capacities.

However, we ought to be careful not to overstate the scope of such results. They can only be obtained under special circumstances, e.g., in situations where subjects are prompted to follow instructions that have been carefully selected by an experimenter to elicit certain meta-logical intuitions. Stenning and Yule's experiment does not show that competent reasoners must always access sophisticated meta-logical knowledge whenever they solve a syllogism on paper. More generally, it does not follow from the fact that some lower-level cognitive mechanism is cognitively *penetrable* (under certain conditions) by higher-level knowledge that it is always *penetrated* by higher-level inferences. It is a thoroughly empirical question how much meta-representational and meta-logical knowledge is causally implicated in a given episode of deductive reasoning – a question that we cannot judge from the armchair. Hence I should properly qualify my argument for the HEC. It is based on Hutchins' (1995a) suggestion to "posit the minimum internal structure required to get the tasks done" (ibid., p. 132). This minimalist working assumption, which is meant as an antidote to the overattribution fallacy, is a methodological version of Occam's razor. Like all methodological principles, it is empirically defeasible in the face of evidence that it doesn't work. But its falsification requires that we really have a detailed grasp of "the task" at hand, and do not lean on preconceived notions about what syllogistic reasoners must know or need not know about their task. As I shall now argue, it can be quite tricky to determine how much meta-representational insight into the problem space of a given domain is necessary to qualify as a competent problem-solver in that domain.

Suppose your friend invites you to play with him a few rounds of the "Game of 15."[66] It is a game for two players who take turns picking numbers (without repetition) from the pool of 1, 2, 3, 4, 5, 6, 7, 8, and 9. The first player to have selected three numbers that exactly add up to 15 wins the game. In the beginning, you always lose to your friend – you find it hard to consider all the rele-

---

66 I borrow this example from Don Norman (2002, p. 126).

vant possibilities, adding up the sums that would result from each choice, while factoring in all the possible responses of your opponent. But one day, somebody teaches you how to arrange the nine numbers in a pattern that effectively turns the Game of 15 into a game of Tic-Tac-Toe (Figure 17). From that point on, you never again lose to your friend.

8	1	6
3	5	7
4	9	2

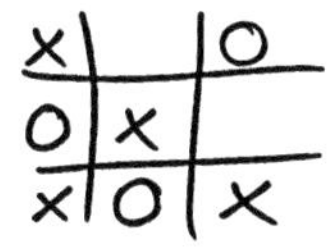

Figure 17. A structural equivalence between the Game of 15 and Tic-Tac-Toe.

Why do we find it so much easier to find a winning strategy in Tic-Tac-Toe than in the Game of 15, even though the two are structurally equivalent? From a purely formal perspective, the decision trees of both games are equally wide and deep. In each case, there are nine opening possibilities, eight possible replies, seven possible moves to start the second round, etc. By the time the beginning player makes her third move (i.e., the first possible winning scenario), there are already 15,120 possible sequences to be considered. It is far too difficult to search through all possibilities in one's head, and would be extremely tedious to do on paper. But by re-representing the problem space in a different format, we dramatically change the type of skills that we can bring to bear on the task at hand. Once we map the numerical structure of the Game of 15 onto the spatial structure of a 3×3 grid on which we have to arrange just two types of tokens (X's and O's), finding a strategy that guarantees at least a draw is almost trivial. We can now take full advantage of our visuo-spatial abilities for recognizing spatial symmetries, and all it takes is to memorize a few basic opening patterns and response strategies such as "If X plays the first move to the center square, O a corner square; then block and draw." Re-representing the Game of 15 as a game of Tic-Tac-Toe transforms the decision-tree into a structure that is both narrow (i.e., having few choices to consider at any given point) and shallow (i.e., with little or no need to plan ahead). This turns the problem of selecting a number for one's next move into a greatly simplified pattern-completion task which requires far less cognitive effort (Gigerenzer, Todd, et al., 1999; Gigerenzer, 2000; Zhu & Gigerenzer, 2006).

Now let's return to our meta-theoretical problem of domain-identity. Once I start using the translation map in Figure 17, have I ceased to play the Game of 15? The answer depends on which process we are considering. Following Hutchins, we have to distinguish between the task performed by the brain (or the person), and the task that is accomplished by the person+artifact system. The process which accomplishes the former now lies in the domain of Tic-Tac-Toe, whereas the process which accomplishes the latter still counts as playing the Game of 15. If I can re-frame the problem space of the Game of 15 as a pattern of Tic-Tac-Toe, my brain doesn't have to represent and work through a complex numerical decision tree. I can directly search for a winning strategy in my novel task representation by treating the O's as *naughts* (rather than as numbers) and the X's as *crosses* (rather than as numbers).

In response, the internalist is going to argue that this translation procedure works only because the subject has acquired the necessary meta-representational competence. In order to win a Game of 15, it is not enough to be able to search in the space of Tic-Tac-Toe. I also need to be able to translate individual positions and moves between the two external representational schemes. But that requires that I recognize the X's and O's as *numbers* (rather than as crosses and naughts), which draws on my meta-representational knowledge that there is a structural equivalence between the two schemes. Doesn't this leave us with the original objection still unanswered?

At this point, we ought to make explicit an important assumption made by the internalist. This assumption is that all meta-representational cognition must go inside the head. But in the example we have sketched, a person's meta-representational cognizance of the structural equivalence between the Game of 15 and Tic-Tac-Toe need not be an entirely internal affair. It is partly constituted by the translation manual represented in Figure 17, which is openly available for a player's visual inspection. It is certainly possible to memorize the chart, if the need arises. But unless we artificially constrain the task (e.g., by introducing certain regulations of play), a player can simply consult a copy of the external chart before deciding on her move. Based on the information in the chart, she can simply translate individual tokens on paper, one at a time (Ballard et al., 1997), without ever "internalizing" the whole chart. Viewed from the perspective of the brain, this is again a relatively shallow, perceptual pattern-matching task which does not require the use of sophisticated internal knowledge about the underlying problem space. You don't have to be a math wizard to become a good Tic-Tac-Toe player, and if you have perceptual access to an accurate chart, you don't have to become one to notice the structural correspondence between the Game of 15 and the game of Tic-Tac-Toe.

In sum, these considerations show that the objection commits another instance of the "overattribution fallacy." The appropriate reply is therefore to reiterate our "division of labor" hypothesis between internal and external resources at the meta-level. *Da capo al fine*: just as syllogistic reasoning is distributed over internal and external resources, so is our meta-representational knowledge of syllogistic reasoning.

5.5. The "coupling-constitution" fallacy that wasn't

Let me now summarize my line of reasoning in this chapter. My primary goal has been to construct a "tie-breaker" in favor of the claim that cognition is sometimes literally *extended* (HEC), as opposed to the competing hypothesis of *embedded* cognition (HEMC; cf. Chapter 1.4). In favor of the HEC, I have presented a representational argument that is framed as a meta-theoretical problem about how to determine the domain-identity of a cognitive process. As a concrete working example, I described a situation which requires that we establish the cognitive type-identity of two different token-processes of syllogistic reasoning. Both processes involve the manipulation of external representation systems. In one case, a person uses the diagrammatic method of Euler Circles; in the other case, a different person uses the sentential method of Natural Deduction. The challenge I raised was to say on what grounds we are entitled to consider both token-processes as episodes of *syllogistic reasoning*. In response to my challenge, I proposed a "division of cognitive labor" hypothesis. On the one hand, external representations are dependent on perceptually grounded representational mechanisms inside the brain to have a causal influence on our behavior. On the other hand, the internal representations on which the neural mechanism operates are dependent on our use of external representations to have their syllogistic content fixed. Thus if we individuate cognitive processes in terms of the distinctive, non-derived content of the mental representations on which they operate, it is only the functional integration of internal and external representational vehicles which qualifies the extended computational processes as two episodes of *syllogistic reasoning* in good standing. Therefore, we ought to embrace the HEC, rather than its conservative internalist rival, the HEMC.

What other implications does my argument have for the current debate over extended cognition? Clearly, a major point of contention in this debate has been the question of whether the causal intercourse between brain, body, and environment is intimate enough to spawn the existence of extended cognitive processes, and/or extended cognitive systems. According to Adams and Aizawa (2010), the "coupling-constitution fallacy" is indeed "the most common mistake extended mind theorists make" (ibid, p. 68) when they tacitly move "from the observation that a process X is in some way causally connected (coupled) to a cognitive process Y to the conclusion that X is part of the cognitive process Y" (ibid., 2009, p. 81; see also Adams & Aizawa, 2008, Chapters 6-7). As a non-cognitive example where said fallacy should be apparent, they point out that the expansion and contraction of the bimetallic strip in a thermostat is closely coupled to the ambient temperature of a room, yet this gives us no reason to consider the expanding and contracting strip as extending into the room (Adams & Aizawa, 2009, loc. cit.).

A similar criticism is made by Rupert (2004, pp. 395-6), who argues that even if one grants that we cannot fully comprehend many episodes of human cognition unless we consider the wider context in which they occur, it does not follow that the embedding context is therefore a part of cognition. For instance, to understand the historical significance of Nazi Germany's invasion of Poland, one ought to know a great deal about the economic conditions in Germany during the 1920s, but this does not make those economic conditions (mereological) parts of the invasion. What the extended mind theorist would need, Rupert suggests, is a further premise that would help to close the suggested inferential gap: the principle that whenever an understanding of A's relation to B is significantly relevant to our understanding of A, we should posit a single system, A-B, as a single unit of study (ibid.). But such a premise (the "A-B principle") seems awfully strong. If taken literally, it would excessively multiply the number of cognitive systems, most of which do not persist beyond very short periods during which an organism interacts with various bits and pieces of its current environment.[67]

Is my argument guilty of committing a coupling-constitution fallacy? Let us first closely attend to the putative form of the fallacy. According to the critics, the supposedly flawed inference is supposed to go as follows:

1. Process X is in some privileged way causally coupled to process Y

Therefore, process X is a (mereological) part of Y.

But this is not the form of the argument I have presented, nor is it generally the form of the most prominent arguments for the HEC. For instance, when Clark and Chalmers (1998) claim that a notebook (X) is in some privileged way causally coupled to Otto-the-cognizing-organism (Y), they do not wish to assert that the notebook is part of Otto-the-cognizing-organism. Instead, what they assert is that if X causally interacts with Y in some principled fashion, the functional integration of X and Y can give rise to an extended cognitive system Z (e.g., an extended memory system) which encompasses both X and Y among its proper parts. Hence none of the above examples is particularly apt to bring out what I take to be the crux of the matter: when do we have good reasons to posit such an integrated process/system Z, and what role does Z play in cognitive-scientific explanation?[68]

---

67 See Chapter 6.2 for further discussion of this objection.

68 All of the above examples ought to be re-phrased correspondingly, e.g., as: Is there an extended temperature control system which subsumes elements from inside and outside

We cannot hope to ascertain the epistemic value of HEC-style explanations – and the systemic taxonomies which they presuppose – on apriori grounds. Thus we agree with Rupert (ibid.) that it is counterproductive to endorse an unrestricted version of the "A-B Principle." This would quickly lead either to a gratuitous proliferation of extremely short-lived cognitive systems, or perhaps the monist conclusion that there is really only one huge cognitive "mind-meld," assuming that everything is somehow contextually dependent on everything else. Although we should not outright dismiss either possibility on the spot, the onus here is on the HEC-theorist to tie the occurrence of extended cognition to specific and sufficiently robust forms of brain-body-environment interactions.

From this perspective, our representational argument can be viewed as the careful description of a specific type of situation in which we have good reason to apply the A-B Principle. The situation involves a systemic coupling between a person and an external notation that is sufficiently (i) causally intimate and (ii) semantically constitutive so as to treat them both as proper parts of a hybrid reasoning system. The functional benefits of the integrated coupling between person and artifact can be characterized from the two perspectives which Norman (1991) has called the *systemic view* and the *personal view* (see Chapter 4.2). From the systemic (person+artifact) view, the hybrid reasoning system has superior inferential capabilities that are not possessed by any of its components. From the personal view, the most salient feature of the notational system is that it transforms the task of deductive reasoning for the person, allowing her to co-opt various pre-existing (e.g., visuo-spatial) skills and background knowledge. But this transformation also adds new functionality to a person's cognitive repertoire. For instance, in virtue of the way in which Euler Circles are designed, a competent EC-reasoner can draw syllogistic inferences *by means of* manipulating circles and crosses.

But the debate over the "coupling-constitution" fallacy doesn't end here. The critics have also complained about a more subtle version of the underlying error, which Adams & Aizawa have termed the "systemic" version of the coupling-constitution fallacy (Adams & Aizawa, 2008, Chapter 7). It differs from the simple version by its extra appeal to the notion of a *cognitive system*. The first step of the criticized argument moves from the premise that there is an integrative coupling between brain, body, and environment to the conclusion that there is an extended cognitive *system*. From there, it moves on the conclusion that cognitive *states* and *processes* extend beyond the head into the world. In trying

the thermostat among its proper parts? Is there a historical process which encompasses both the economic conditions of the Weimar Republic and Nazi's Germany's invasion of Poland?

to expose the fallacious nature of this two-step argument, Adams and Aizawa provisionally grant the first inference, and then challenge the second inference with the observation that "it does not follow from the fact that one has an X system that every component of the system does X" (2009, p. 84). For instance, the components of a typical air-conditioning system include a thermostat, a compressor, an evaporation coil, and a fan. Within this system, the process of the air cooling as it passes over the evaporation coils is of the air-conditioning type. However, it would be misleading to subsume the liquefaction of Freon or the electric processes unfolding within the circuit of the thermostat under the same description as an "air-conditioning" process. Is my argument guilty of committing this kind of fallacy?

In response, I should first point out that Adams and Aizawa's machine analogy is quite helpful to connect our computational-representational argument for the HEC to the broader conception of *mechanistic explanation* in cognitive science (Craver & Bechtel, 2006; Bechtel, 2008). What do we mean by a mechanistic explanation of a cognitive process? First, we identify a complex system as a machine of type S by the task that it ought to perform. For instance, by analyzing something as a (syllogistic) reasoning system S, we automatically highlight a range of subtasks which the mechanism M that underlies the operation of S ought to be able to perform (e.g., discern valid from invalid arguments, derive valid conclusions, etc.). Second, we identify the relevant components of S (and their properties) in terms of their causal relevance for the operation of M. If a system is modular, we can break down the functions of M into a set of more basic functions, while we concurrently break down S into its structural components, in order to figure out what each of the components does (e.g., representing the major premise of an argument, detecting critical individuals in valid inferences, etc.). Finally, we describe the mode of organization between the components that is required for M to work. Hence it is a fairly natural consequence of mechanistic explanation that the various components of a complex information-processing system of X typically do not all perform functions of type X. But why should this observation undermine our analysis of syllogistic reasoning as an extended reasoning process?

First, when we consider an extended cognitive system S as a machine for syllogistic reasoning, it does not follow that every part of S – or *any* proper part of S, for that matter –is recruited for the task vis-à-vis the global function that we properly ascribe to S. For instance, in order to establish Premise (4), we have argued that for HOLIST, the within-the-head part of M1 is not, properly speaking, *syllogistic* at all, but involves the ability to produce and understand various geometrical configurations of circles. Similarly, following the heuristic that "If X plays the first move to the center square, O a corner square; then block and draw" yields a strategy that works well for a player who already understands the basics of Tic-Tac-Toe. But considered in isolation, e.g., without taking into account a person's knowledge of the correspondence between Tic-Tac-Toe and the Game of 15, the heuristic would be useless. It would not have any function at all for playing the Game of 15. Second, it is also not true that our argument haphazardly lumps together any aspect of the causal interactions between a person and an external representation system under the generic description "reasoning process." For instance, our argument does not imply that the exact way in which a person moves her arm has an intrinsically syllogistic significance, even though moving one's arm is causally implicated in the manipulation of external representations. The reason we can draw such a distinction is that our argument for the HEC embraces a strongly representational criterion for cognitive domain-identity that is based on the non-derived content of mental representations. This should be music to the ears of Adams and Aizawa (2001, 2008, 2010), who hold that non-derived intentionality is an essential "mark of the cognitive" (see Chapter 2.10). I thus conclude that our argument for the HEC is immune to Adams and Aizawa's criticism, because it does not have any of the allegedly flawed implications which they presuppose.

5.6. Turning the parity principle inside-out

I would like to end my discussion of syllogistic reasoning with a critical note about Clark and Chalmers' (1998) advocacy of the principle of parity (Chapter 2). While there has been considerable debate over the role of parity arguments for extended cognition, comparatively little attention has been paid to the "naturalist" bias that its original formulation seems to imply. The parity principle trades on the assumption that cognitive processes which are going on inside the brain always provide the gold-standard which sets the bar for any other processes that also ought to count as cognitive. However, many forms of advanced cognition would be so taxing to our biological brains that they are practically never carried out inside the head.[69] For most of us, the acquired mastery of sophisticated, notation-driven forms of deductive reasoning would seem to fall into that category. Second, many paradigmatic cognitive operations which are far too complex to be entrusted to single individuals – even equipped with artifacts – are almost invariably carried out by teams, organizations, or entire institutions. Examples include the complex decision-making processes of large corporations, the institutional memory of the legal system, and large-scale net-

---

69 Perhaps this limitation is a direct result of certain psychological or neurobiological constraints that are nomologically necessary, but we need not make that stronger claim here.

works of scientific collaboration (Gallagher & Crisafi, 2009; Theiner, Allen & Golstone, 2010).

On the other hand, it is also true that we can – with a lot of practice and dedication – internalize certain aspects of representational formalisms which begin their life on paper. When this happens, internalized representational schemes can provide a kind of "virtual cognitive architecture" that runs on top of our biologically more basic processing capabilities (Dennett, 1991a; Dehaene, 1997; cf. Chapter 4.3). One could go so far as to say that internalized cognitive architectures form an interesting sub-category of cognitive extensions that "paradoxically enter the traditional stronghold of internalism, the brain" (Sutton, 2006, p. 239). As an example, Sutton (ibid.) mentions the complex memory architectures acquired by baroque monks and scholars (Carruthers, 1998); we can add the baffling ability to play blindfold chess (Hearst & Knotts, 2008) to this sub-category.

Much like virtuoso displays of blindfold chess, completely internalized executions of an Euler Circle algorithm (relying purely on visually imagined representations of EC) are hard to learn and remain prone to error; they are certainly the exception rather than the rule. But let us suppose that an exceptional member of the population of HOLISTS becomes expertly capable of doing all the diagrammatic calculations in her head. If our argument is sound, doesn't it entail that those internalized surrogate performances would *not* be "the real McCoy," i.e., would not qualify as episodes of *syllogistic reasoning* in good standing? This is because their syllogistic content would be merely derived from the intentional content of external representations that were originally manipulated on paper. I think this would be an implausible consequence of my argument. But in order to block it, we must allow that the principle of parity can also be turned inside-out:

INSIDE-OUT-PARITY: If, as we confront some task, a part of the *brain* functions as a process which, were it done using *environmental* resources, we would accept as part of the cognitive process, then that part of the brain is part of the cognitive process.

6. Three objections to the extended mind rebutted

6.1. Three conceptions of the extended mind thesis

If cognition extends beyond the skin into the world, who are the proprietary subjects of such extended cognitive states and processes? Does the occurrence of extended cognition make plausible the existence of extended *minds* – or perhaps even extended *selves*, as Clark and Chalmers (1998) have argued:

> "What, finally, of the self? Does the extended mind imply an extended self? It seems so. Most of us already accept that the self outstrips the boundaries of consciousness; my dispositional beliefs, for example, constitute in some deep sense part of who I am. If so, then these boundaries may also fall beyond the skin. The information in Otto's notebook, for example, is a central part of his identity as a cognitive agent. What this comes to is that Otto *himself* is best regarded as an extended system, a coupling of biological organism and external resources. To consistently resist this conclusion, we would have to shrink the self into a mere bundle of occurrent states, severely threatening its deep psychological continuity. Far better to take the broader view, and see agents themselves as spread into the world" (ibid, p. 18).

Let us call "extended self" thesis the claim that the self extends beyond the boundaries of the biological individual. In the passage at hand, Clark and Chalmers seem to endorse the extended self thesis, but their endorsement is limited by the fact that they do not tell us what exactly they take to be a self. The notion of a *self* is notoriously ambiguous. When people ordinarily talk of "the self," they tend to conflate several different attributes that we commonly associate with the human mind. One of the attributes of selfhood is the experience of *agency*. We think of ourselves as agents who can do things in the world – make decisions, form intentions, act upon those intentions, and exercise causal control over our actions. Closely related to the notion of agency is the idea that selves are subject to praise and blame. When we praise or blame somebody for their actions, it means that we hold them responsible for actions which we take to be under the agent's self-control. We also believe that selves are relatively stable and persisting *persons* who retain their numerical identity over time. For instance, when we celebrate a person's tenth birthday, we commemorate an event which happened ten years ago to somebody who we take to be identical with the person who we are now celebrating, despite having undergone significant physical and psychological changes. Finally, we associate with selves the capacity to have phenomenally *conscious* mental states. The self-intimating nature of consciousness underpins the related idea that minds enjoy some form of privileged epistemic "first-person" access to their own conscious mental states. In sum, there are a number of distinct attributes – *agency*, *personal identity*, *consciousness,* and *first-person access* (and the list could no doubt be continued) – all of which go into the metaphysical hodgepodge which is our folk-psychological notion of "the self."

In his writings, Andy Clark has occasionally flirted with the extended self thesis, but without ever fully embracing a "thicker" notion of an extended self.[70] This may be due to two main reasons. First, the fate of the "extended self" thesis is likely to hinge on the formative role of the phenomenology of agency, as well as the related sense of bodily ownership (Gallagher, 2000; de Vignemont, 2007; Tsakiris, Bosbach, & Gallagher, 2007). If the physical machinery that is constitutive of our phenomenally conscious experience of agency and bodily ownership is wholly located inside the body (cf. Clark, 2009), it would seem problematic to infer the existence of extended selves from weaker versions of the HEC. A second reason for Clark's reluctance may be his belief that our vernacular idea of a person and its associated notions of agency and responsibility are ultimately – as Locke (1694/1975) would have put it – "forensic" notions, i.e., notions whose application is more a matter of legal convention than of metaphysical substance (Clark, 2005a, p. 10; Clark, 2006b). For Clark, locating persons in the physical world is essentially a matter of identifying "rational systems" (Rovane, 1998), i.e., a theoretically flexible, interpretive practice that is likely to adapt as we become increasingly connected to our cognitive technologies (Clark, 2003). Hence it remains to be seen whether the extended self thesis (over and above the HEC) has much to contribute to the causal-explanatory practices of cognitive science. But looking beyond cognitive science, it is certainly an interesting project to explore stronger versions of the extended self thesis which view the human self as being culturally enacted across the skin barrier, actively incorporating tokens of our material culture (Malafouris, 2008).

But even if we steer clear of the muddy waters surrounding our ordinary talk of selves, there are still two importantly different ways of articulating the idea of extended cognition. They differ with respect to who we ought to consider as proprietary *bearers* or *subjects* of extended cognitive properties.[71] According to the more common interpretation, which is evidently the one favored by Clark and Chalmers (1998), the subjects of extended cognitive properties are (at least typically) no less extended than the entire physical machinery on which their manifestation metaphysically depends. I shall refer to this as the "wide subjects" conception of the extended mind thesis. For instance, if the inscriptions in Otto's

---

70 See e.g., Clark, (1997 pp. 216-18; 1998d, p. 367), and Clark & Chalmers (1998, p. 18).

71 I am using the term "property" broadly as a generic metaphysical placeholder for states, events, processes, capacities, and dispositions.

notebook are metaphysically constitutive vehicles of extended beliefs, the "wide subjects" view implies that we ought to treat them as dispositional beliefs *of* an extended cognitive subject (i.e., *of* Otto-plus-his-notebook).

However, there is also a different and slightly more conservative way of characterizing the subject of extended cognition. This alternative view endorses the HEC as a claim about cognitive properties, but at the same time retains the idea that it is typically the biologically-bound individual who is the relevant subject of those properties. Accordingly, we ought to treat the entries in Otto's notebook as extended beliefs *of* Otto-the-organism-bound-subject. This alternative interpretation, which I shall dub the "narrow subjects, wide realizations" conception of the extended mind thesis, has been developed in great detail by Robert Wilson (see especially Wilson, 1995, 2001, 2004). The ontological import of the "narrow subjects, wide properties" conception goes well beyond the HEC. Wilson has argued for the broader metaphysical thesis that a significant portion of higher-level properties (e.g., biological fitness) has so-called *wide realizations* which stretch beyond the spatiotemporal boundaries of their bearers.

Internalist critics have argued that both conceptions of the extended mind thesis have profoundly counterintuitive if not metaphysically incoherent implications, and thus ought to be rejected. In this chapter, I shall discuss several objections of this sort that have been raised by Robert Rupert (2004, 2006, 2009, 2010). Against Clark's "wide subjects" view, Rupert has argued that the facultative and short-lived nature of most extended cognitive systems would jettison the genuine subject matter of psychological theorizing: the persisting individual who can exercise her cognitive capacities across a variety of different environmental contexts. Against Wilson's "narrow subjects, wide realizations" conception, Rupert has argued that the existence of cognitive properties with wide realizations is metaphysically incoherent, because it would mean (*per impossibilem*) that certain environmental states would be both causes and realizers of cognitive states. The goal of this chapter is to scrutinize Rupert's objections, which provides us with a great opportunity to explore the metaphysics of the extended mind. I shall argue that both conceptions of the extended mind thesis survive Rupert's criticisms unscathed.

6.2. The argument from integration

Against the wide-subjects version of the HEC, Rupert has argued that it would lead to an unmotivated proliferation of relatively short-lived, "soft-assembled" cognitive systems which do not coherently persist over time and across different environments (e.g., Rupert 2004, pp. 425-428, pp. 49-52; 2006, pp. 2-10; 2009, Chapter 3; 2010). This implication would not only violate our folk-psychological intuitions about how we ordinarily think about the existence of minds, but also jettisons the subject matter of scientific psychology: the behavior of persisting, relatively unified, and organismically bounded cognitive systems. As he puts it, the wide-subjects conception of the HEC "introduces a profligate set of distinct cognitive systems the richness of which confounds standard methodology with no productive replacement in sight" (2006, p. 10).

To illustrate his worry, Rupert draws attention to various areas of cognitive psychology that are committed, or so he argues, to an organism-bound view of cognition. For instance, in educational psychology, our pedagogical efforts are aimed at the acquisition of cognitive skills that a student can exercise in a variety of circumstances, and beyond the learning environment in which she was originally instructed: "Neither teacher, tutor, computer, chalkboard, nor textbook can follow the child around for years to follow the lesson" (ibid., p. 6). Experimental psychologists assume that the cognitive subject which undergoes pre-experimental practice sessions is the same subject that later participates in the actual experiment (ibid, p. 6.). Even developmental psychologists, whose goal is to study how certain cognitive capacities change over time, are interested in change only insofar as it affects a "single developing human" (ibid., p. 5). According to Rupert, the best explanation of the recurring behavioral patterns that arise from the causal interactions between an organism and its environment "appeals to reappearance of the organismic [cognitive] system with its persisting properties; these include higher-order properties of being likely to change first-order properties in certain regular ways as the result of experience or aging" (Rupert, 2009, p. 38).

In contrast to the persisting collection of organism-bound cognitive capacities, the vast majority of cognitive systems posited by the HEC-theorist are highly facultative (i.e., easily decoupled) and very short-lived in their existence. When somebody writes a scholarly paper, "the author couples with notes, then takes a break, goes for a walk, perhaps thinking about the notes as she ambles about campus. A year later, the notes are, more likely than not, in a trash bin or filed away" (Rupert, 2006, p. 4). When perceiving a scene, "the subject might use the bedroom as its own best perceptual model, but once the subject leaves the house, there is no reason to think the subject-bedroom system persists" (ibid, p. 4). Similarly, if it weren't for the persistence of individual cognitive subjects, what reason would a developmental psychologist have to study the temporal succession of distinct extended cognitive systems such as "Sally, aged two, together with a ball she was bouncing on some particular day, Johnny, aged five, together with the book he was reading on some particular afternoon, and Terry, aged seven, plus the stimulus item he has just been shown by an experimenter" (ibid., p. 5). On all these occasions, psychologists attend to individual tokens of cognitive behavior only because they are eager to draw inferences about the persisting (or perhaps developing) cognitive capacities of individual subjects. In order to discover law-like generalizations, "we must identify the capacities involved, their distinctive characteristics, and the supporting mechanisms, the operation of which explain single instances of behavior (2009, p. 37; see also Wilson, 2002, p. 630). The only way to prevent the looming disintegration of individual minds into a fleeting bundle of barely overlapping cognitive systems – and thus to preserve the methodological integrity of psychology – is to lock up the cognitive subject inside the organism.

In his recent monograph, Rupert (2009, Chapter 3) has outlined his own positive conception of what constitutes a cognitive system. He calls it the "systems-based" view of cognition. Put succinctly, his view asserts that a state (or process) is cognitive just in case it is part of an *integrated cognitive system*. For Rupert, it is the boundary of an integrated cognitive system which marks the only plausible criterion for separating what merely causally contributes to the production of cognitive phenomena (e.g., an external tool), from what is distinctively cognitive. Based on his demarcation criterion, Rupert argues that external resources are typically not *constitutive* parts of human cognition, but only *used* by the cognitive system, because they are not effectively integrated into the human cognitive system. The systems-based view yields a direct argument against the HEC that I shall dub the *argument from integration* (cf. Rupert 2009, p. 45; 2010):

1. The empirical success of "orthodox" cognitive-scientific explanations requires a collection of capacities, mechanisms, skills, and abilities that persist across different environmental contexts.
2. The presence of this collection is best explained by the persistence of an integrated cognitive system.
3. A state or process is cognitive just in case it is part of an integrated cognitive system.
4. For normal human beings, the integrated cognitive system is realized within the biological organism.

Therefore, the HEC is false.

Let us now carefully scrutinize Rupert's integration argument.

6.2.1. Up and down the method of functional analysis

Premise (1) of Rupert's integration argument draws attention to a standard explanatory practice which is said to involve the identification of cognitive capacities (e.g., categorization, planning, perception, memory, language-use, or inference), their characteristic features, and the supporting mechanisms whose operation causally explains token manifestations of those capacities on a given occasion (Rupert, 2009, p. 37). In "orthodox" cognitive science, a commonly used species of this general explanatory strategy is known as the method of functional analysis (Dennett, 1978; Marr, 1982; Cummins, 1983; Lycan, 1987). According to Cummins (1983), the method of functional analysis can be characterized as follows. The basic strategy (stage 1) is to analyze the behavior of a cognitive system as computing a function of interest (or, to put it more broadly, as solving an information-processing problem). To explain how this is done, we break down the global function-to-be-computed into an organized collection of simpler functions, each of which can be computed by simpler sub-processors (stage 2). This procedure can be iterated until we reach a bottom level of primitive functions (stage 3) which can be directly carried out by localizable parts whose operation we can explain directly in mechanical or neurobiological terms. As we have pointed out earlier (Chapter 3.4), the method of functional analysis is a particular instance of the more general family of "decomposition and localization" strategies for explaining the behavior of complex systems (Bechtel & Richardson, 1993).

Rupert is right that the method of functional analysis has been a cornerstone of orthodox cognitive science. However, the central features of that explanatory practice are by no means restricted to the cognitive analysis of biological organisms. Below the level of the organism, it has been gainfully employed to explain cognitive properties of the brain, specific neural regions within the brain, or even dedicated neural circuits within particular brain regions (Craver, 2007; Bechtel, 2008). Above the level of the biological organism, the method of functional analysis has been used to analyze the cognitive capacities of supra-individual information-processing systems that comprise the socially organized interactions between people and artifacts. For instance, it has been used to study processes of collective memory, learning, and decision-making in social organizations (Sandelands & Stablein, 1987; Walsh & Ungson, 1991; Argote, 1999). Proponents of the *distributed cognition* framework have shown that we can move up the primary unit of functional analysis from the individual to the collective level in order to study the dynamics of collaborative work activities (Hutchins, 1995a; Hollan et al., 2000). The explanatory targets of the distributed-cognition approach are heterogeneous information-processing systems that are "soft-assembled" from people, artifacts, internal and external representations, and operate under rich social and organizational constraints (see Chapter 1.3 for discussion). It has been used to study the functioning of ship navigation teams (Hutchins, 1995a), cockpit crews (Hutchins, 1995b), software teams (Flor & Hutchins, 1992), call centers (Ackerman & Halverson, 1998), control centers (Garbis & Waern, 1999), and even the Elizabethan theatre (Tribble, 2005). In each case, our choice of the relevant level of explanation depends on our prior decision as to what type of behavior we want to study, and hence which system we have to isolate; the cognitive behavior of persisting organisms is only *one* of those levels. This fact alone should be enough to raise doubts about Rupert's attempt to derive substantial conclusions about the boundaries of the mind from the commitment to a capacity-centered methodology.

Historically speaking, it is no coincidence that we can apply the method of functional analysis to social systems with relative ease. Functional analogies between persons and corporate entities have long influenced our conceptions of the human mind, as Austen Clark (2004, esp. pp. 410-25) has pointed out. For instance, Plato's *Republic* is modeled on the soul; Hobbes' *Leviathan* is an artificial man. Hume picks up the Platonic metaphor when he says that

> "[...] I cannot compare the soul more properly to anything than to a republic or common wealth, in which the several members are united by the reciprocal ties of government and subordination, and give rise to other persons, who propagate the same republic in the incessant changes of its parts" (Hume, 1975 p. 261; cited after Austen Clark, 1994, p. 410).

Clark (ibid, pp. 410-15) reminds us that for contemporary functionalists, the suggested functional commonalities are considered to be more than merely metaphorical. If functionalism is true, minds *are* corporate entities of sorts, insofar as mental states are analyzed relationally, in terms of their "job description" within the organization of a cognitive system (Putnam, 1975a; Block, 1978). Clark (ibid.) also notes that Dennett's once-optimistic characterization of "homuncular" functionalism was that "[i]f one can get a team or committee of *relatively* ignorant, narrowminded, blind homunculi to produce the intelligent behavior of the whole, this is progress" (Dennett, 1978, p. 123). In William Lycan's (1981) words, functionalism envisions the mind of a person as a kind of corporate entity,

> "an integrated system of intercommunicating 'departments' that cooperatively go about the business of interpreting the stimuli that impinge on the corporate organism and of producing appropriate behavioral responses. In this model, a psychological description of a human being will consist of a set of flow charts, nested hierarchically. [...] The subsystems will be characterized in terms of their respective corporate responsibilities: 'executive unit,' 'perceptual analyzer,' 'speech center,' 'buffer memory,' and the like" (ibid., p. 140).

As a final comment on Premise (1), it is also important to bear in mind that not all extended cognitive systems are equally "fleeting" in nature, as Rupert seems to suggest. Within the two-dimensional framework proposed by Wilson and Clark (2009), temporal duration is one of the dimensions on which cognitive extensions vary on a continuum between (i) ephemeral ("one-off") add-ons, to (ii) temporary, but repeatable hybrid ensembles, all the way up to (iii) relatively permanent cognitive augmentations (e.g., Otto's notebook) that also score highly on a number of "coupling conditions" (Clark & Chalmers, 1998; see Chapter 2.8). An example of a relatively permanent alliance between an organism and a bio-external cognitive resource is the extended cognitive system formed by Kanzi and his symbol pad (Wilson, 2004, Chapter 8; Wilson & Clark, 2009, p. 67). Kanzi is a human-raised bonobo (pygmy chimpanzee) who grew up in a human-centered environment on the research grounds of Sue Savage-Rumbaugh (Savage-Rumbaugh & Lewin, 1994; Segerdahl et al., 2005). As an integral part of his cognitive development, Kanzi has regularly interacted with a portable, oversized (and now computer-monitored) keyboard which has a large collection of lexigrams and symbolic icons printed on it. Kanzi's prolonged exposure to an external, perceptually stable symbol system has not only significantly affected his inner cognitive architecture, but Kanzi's manipulations of the keyboard are in many ways functionally indispensable ingredients of an impressive range of cognitive skills that are unrivalled by his keyboard-deprived biological conspecifics. Even though the soft-assembled cognitive system which comprises the interactions between Kanzu and his keyboard is primarily task-oriented, and thus more facultative than the permanent biological integration of Kanzi's body parts, the reliable presence of the keyboard in Kanzi's psychological development is clearly more than ephemeral.

6.2.2. A dilemma for Rupert's mark of the cognitive

In what follows, I want to focus on Premise (3), i.e., Rupert's systemic mark of the cognitive, and examine its compatibility with Premises (2) and (4) of the integration argument. Rupert defines an integrated cognitive system (a "cognitive architecture") as a cluster of highly interdependent mechanisms, frequently working together in overlapping ways in the production of a wide range of cognitive phenomena, across a variety of conditions (2009, pp. 41-43). Only mechanisms which form part of an integrated cognitive system are literally cognitive; other resources are merely *used* by the cognitive system. Rupert proposes his demarcation principle as a theoretical hypothesis in cognitive science, not a conceptual analysis of what it *means* to be a cognitive system. Hence it cannot be refuted "from the armchair," as it were, by providing purely imaginary counterexamples. For Rupert's criterion to do the work which he intends it to do, it ought to jointly satisfy two requirements. First, it must be narrow enough to exclude the deployment of external resources as cognitive, consistent with Premise (4). Second, it must be wide enough to let in all of the organism-bound cognitive machinery which "orthodox" psychology recognizes, consistent with Premise (2).

To cope with the first requirement, Rupert adds a "perceptual screening-off" condition to exclude commonly available external resources (e.g., the sun) which earn a high integration score simply by their causal effects on various "input-level" mechanisms (ibid., p. 43). For instance, the sun co-contributes to a wide variety of cognitive outcomes, but does so only by stimulating retinal cells. However, this condition begs the question against the HEC in multiple ways. First, it invokes a strict demarcation between cognitive and perceptual mechanisms, a separation which many HEC-friendly accounts deny (Thelen and Smith, 1994; van Gelder, 1998; Hurley, 1998b; Spivey, 2007; Chemero, 2009). Second, it presupposes an account of perception which bears a suspiciously strong resemblance to traditional internalist "input-output" models of perception (Hurley, 1998b). Against this picture, proponents of extended perception have argued that the body's own morphology and activities (such as eye-movements, foveation, or head-turning), but also ambient environmental structures are equal partners of the brain as part of a continuous, boundary-crossing feedback cycle that dynamically links perception to action (Noë, 2004; Matthen, 2005). Third, how we should properly characterize a person's *epistemic actions* (Kirsh and Maglio, 1994) in the world, which for creatures like us are always mediated by sensory channels, is precisely the theoretical dispute in question. For instance, concerning Otto's access to his notebook, what may look like an act of *perception* is really – or so the argument goes – an extended cognitive process of *memory retrieval* (Clark and Chalmers, 1998; Chalmers, 2008, p. xi).

Rupert (2010, p. 351) proposes another fix which adverts to the role of mental representations. He suggests that a mechanism must (at least sometimes) operate on representations with non-derived content to be admitted as a candidate for inclusion in a cognitive system. But there are serious problems with this proposal as well. First, as he points out, it is likely to suffer from circularity. If some representational structure can only be a *mental* representation with non-derived content if it is part of an integrated cognitive system, we cannot appeal to mental representations to individuate cognitive systems. Second, the proposal not only presupposes a representational theory of cognition, but also the somewhat fuzzy distinction between intrinsic and derived intentional content which is not universally accepted even among representationalists. Third, HEC-theorists who accept such a distinction have argued that certain bodily actions (called "deeds" in Rowlands, 2006), but also external representations (e.g., specialized notations, precision instruments) can possess forms of non-derived intentional content if they are properly integrated with the complementary intentional capacities of our biologically based cognitive apparatus (Menary, 2006; Theiner, 2007; Sutton, 2010).

Rupert does not seem to worry that his systemic mark of the cognitive may fall short of the second requirement, perhaps because the image of the skullcap projects a deceptive image of coherence on the internal mind/brain. But it is instructive to apply Rupert's putative mark of the cognitive *inside* the skull. It is a straightforward consequence of Rupert's account that the degree to which a collection of mechanisms forms an integrated cognitive system is inversely proportional to its level of modularity. This is an undesirable implication since many "orthodox" research programs advocate the idea that significant parts of our inner cognitive machinery have a modular structure.

For instance, consider a two-layered cognitive architecture of the type envisioned by Fodor (1983). Let's suppose that the peripheral layer includes a module A for speech recognition (= cognitive outcome a), a module B for face recognition (= b), and a module C for music recognition (= c). In their normal operation, there is no interdependence nor any overlap between A, B, and C. Each module functions autonomously in the production of its proprietary outcome, and does not distinctively contribute to the operation of any other module. Hence A, B, and C are not properly speaking parts of a single cognitive system, but merely non-cognitive resources "used by" the cognitive system, i.e., presumably Fodor's central processing layer. The looming threat of cognitive shrinkage is further compounded if it turns out that even Fodor's central layer has a modular architecture, as claimed by evolutionary psychologists (Cosmides & Tooby, 1994; Pinker, 1997; Buss, 2004). Perhaps Rupert is happy to embrace this conclusion. At one point, he emphasizes flexibility as a central trait of intelligent behavior, and takes it as a theoretical virtue of his account that it captures the distinction between flexibly deployable cognitive mechanisms and "the dedicated, and apparently *unintelligent*, nature of mere tools, reflexes, and canned responses" (2009, p. 44; my italics). However, why should we be inclined to adopt a criterion of the cognitive which is inherently biased against functional segregation and task specificity if those appear to be important organizational principles of the brain (cf. Sporns et al., 2004)?

As a rejoinder, Rupert is going to argue that we should not count how often modules are co-employed in the production of "shallow" cognitive outcomes (such as a, b, c) taken individually, but only in their joint contributions to the "deeper" cognitive outcomes (e.g., belief formation, decision-making) produced further downstream by the central processor. This re-count is bound to raise the integration score of our modules, but only as a rather dull consequence of two related facts: first, that all modules causally contribute to a single central processing system; second, that a broad enough description of cognitive outcomes increases the likelihood for *any* two mechanisms – task-dedicated or not, internal or not – to be co-employed more often. Once we make those features diagnostic of integration, the floodgates open for externally located mechanisms and resources – tools, measuring devices, hand gestures, linguistic artifacts, or iPhone applications – that are also regularly recruited by the central processing system (in tandem with a shifting coalition of internal modules) in a wide variety of outcomes. First, as for internal modules, task-specificity doesn't mean infrequent employment if the work on which one specializes is in sufficiently high demand. Second, human beings actively scaffold their environments (e.g., desks, kitchen space, airline cockpits) to ensure that the necessary resources are ready to be employed on demand (Kirsh, 1995). Third, the use of "special-case"-oriented (rather than general-purpose) routines that work only in benign environments is an adaptive problem-solving strategy if the vast majority of ordinary cases falls into this category (Gigerenzer, Todd, et al., 1999).

We've now come full circle. Applying Rupert's systemic mark of the cognitive in a way that satisfies our second requirement exacerbates the problem raised by our first requirement, and *vice versa*. This tension, I would argue, stems from the fact that what we casually refer to as "the mind" is (probably!) neither like a Swiss Army Knife, nor like a seamlessly integrated cognitive whole. Perhaps a mind is best described as what engineers call a "kludge" – a clumsy agglomeration of parts, haphazardly cobbled-together into an incoherent whole whose operation turns out to be surprisingly efficient. This would certainly seem to be true of extended minds. But if recent evolutionary accounts are correct, the brain itself is the biggest kludge of all, at all levels of organization from regions and circuits all the way down to cells and molecules (Linden, 2007; Marcus, 2008).

Is there perhaps another lesson to be learned from integration-based marks of the cognitive? To see this, let us briefly consider an alternative measure of information integration that was recently proposed to analyze the functional interplay between modularity and integration within the brain (Tononi & Sporns, 2003). In their account, the capacity $\Phi(S)$ of a subset S (of elements of some system X) to integrate information is measured as the minimum amount of causally effective information that can be exchanged across any two complementary parts of S. To receive high values of Φ, a subsystem S must combine heterogeneous connection patterns among different parts of S (specialization) with dense interconnections between functionally segregated parts of S (integration). For instance, the high values of Φ achieved by the thalamocortical system may underpin the formation of an integrated "dynamic core," a promising candidate for the neural basis of consciousness (Tononi, 2004). On the other hand, predominantly segregated networks in brain regions with low values of Φ such as subcortical cycles and loops, but also informationally insulated loops within the cerebral cortex do not form part of the dynamic core, although they are responsible for other, nonconscious forms of cognitive processing.

Now suppose we revise Rupert's systemic mark of the cognitive as follows: x is a cognitive state/process if and only if x is produced and causally sustained by an informationally integrated complex with high values of Φ. If the information-integration account of consciousness is correct, this would demarcate the cognitive *qua* the conscious. Moreover, if the dynamic core which is minimally necessary and also jointly sufficient to produce conscious mental phenomena is entirely located within the boundaries of the thalamocortical system, the suggested demarcation principle would appear to be incompatible with HEC. (Notice that these are two big "IFS".) This would leave us with two options. The first one, recently endorsed by Clark (2009), is to grant that the minimally constitutive supervenience base of phenomenal consciousness is (probably) inside the head, but deny that extended cognition implies extended consciousness. The second option is to endorse the suggested alternative criterion of the cognitive, and thus implicitly shift the main psychological explanandum from the production of intelligent behavior to the production of conscious mental states. Whatever the ultimate merits of such a proposal, it would, at best, be a Pyrrhic victory for critics like Rupert, because a notion of cognition that is essentially tied to consciousness is at odds with current cognitive science in both its internalist and externalist guises.

In sum, I conclude that Rupert's integration argument fails to undermine the HEC. It relies on a systemic principle of demarcation which does not yield an independently plausible account of cognition. In particular, Rupert's criterion is either too restrictive to accommodate the full gamut of organism-bound cognition that is commonly recognized by orthodox research programs, or it is too permissive to rule out extended cognition. Finally, I have briefly probed into an alternative information-theoretic account of cognitive integration, and suggested that the persistence of internalist intuitions about the mental may be more intimately connected to consciousness than orthodox cognitive science would have us believe.

6.3. The argument against "wide realizations"

As we have noted earlier, an alternative "narrow subjects, wide realizations" conception of the HEC has been independently developed by Rob Wilson (1995, 2001, 2004). According to this conception, the physically constitutive realization of certain mental properties extends beyond the biological organism, but those "widely realized" properties can nevertheless be understood as properties *of* organism-bound subjects. Wilson distinguishes the philosophically richer notion of a cognitive *subject*, which he links to the concept of *cognitive ownership*, from the more technical notion of a cognitive *system*. Wilson (2001, pp. 23-25) offers three reasons for closely tying cognitive ownership to the biologically bounded individual.

First, the main target of Wilson's prolonged campaign against Cartesian internalism (Chapter 1.1) has been the assumption that mental states must be physically realized within the boundaries of the individual. For Wilson, however, Cartesian internalism is only a symptom of a deeper-seated metaphysical prejudice that he calls the "standard" view of physical realization (2001). According to this view, the physically constitutive realization of higher-level (e.g., biological, psychological, or social) properties must always be located entirely within the boundaries of their bearers. In order to defeat the "standard" view, it is imperative to show that the material substrate of certain cognitive properties extends into the environment, whereas there are no comparable metaphysical dividends in discarding, in addition, the organism-bound notion of a cognitive *subject*.

Second, there is no denying that the boundaries of the biological organism are usually a reliable heuristic to identify the persisting subjects of mental properties. The persistence of the biological organism marks out persons as "spatio-temporally bounded, relatively cohesive, unified entities that are continuous across space and time" (Wilson, 2001, p. 24). Hence the motivation to posit extended cognitive *subjects* begins to dwindle if the functional coupling with external resources is of an exceedingly fleeting, one-off, and ephemeral nature.

Third, Wilson thinks that common sense rebels against the "wide subjects" view. To illustrate his point, he offers the following analogy. Even though certain mental states (e.g., pain) seem to be completely realized within a particular subsystem of the organism (or the brain), we do not therefore consider these sub-personal systems as the proprietary *subjects* of the mental states which they realize. For instance, "the visual system and its parts can be lesioned, can have imbalances in levels of neurotransmitters, and have certain of its pathways blocked [...]. But it is the *individual* who perceives, who suffers from a visual agnosia, who experiences a hallucination" (ibid., p. 25). In order to accommodate these folk-psychological intuitions, Wilson has refrained from the extra step of positing extended cognitive subjects, opting instead for the somewhat weaker view that only cognitive states and processes are properly speaking extended.

For the purpose of my discussion, I shall not quarrel with Wilson's predilection for "narrow subjects," nor shall I attempt to take sides in the dispute between the two alternative construals of the HEC that I have distinguished. My more modest goal is rather to explore the metaphysical foundations of Wilson's non-standard version of extended functionalism, and defend it against Rupert's criticisms.

6.3.1. The principle of subject-property alignment

Would the argument from integration that we discussed earlier (Chapter 6.2) work equally well against Wilson's "narrow subjects, wide properties" view of the HEC? Robert Rupert (2006, p. 11; 2009, pp. 61-68) has argued that it does, provided that we add a "general metaphysical principle" (2006, p. 11) that links the location of higher-level properties to the location of their lower-level realizers. Here is a concise statement of Rupert's reasoning:

> "psychology requires persisting systems; these persisting systems are organisms; so the boundary of the realizations of those systems' cognitive states should be the boundary of the organism. Perhaps a general metaphysical principle stands behind this line of thought: that the realizer of a state *a* should appear in a portion of the same spatio-temporal region as the subject of state *a*. So long as some viable theory of realization respects this principle, that theory of realization is to be preferred, if for no other reason than to avoid the mystery of realizers' being somewhere other than the location of the states they realize" (Rupert, 2006, p. 11).

For Rupert, the suggested "general metaphysical principle" has the status of a global constraint on the notion of *physical realization* (cf. Rupert, 2009, pp. 61-68). Let us refer to this constraint as the principle of "subject-property alignment" (SPA):

SUBJECT-PROPERTY ALIGNMENT (SPA): Given any higher-level functional property *P*, the physical realizer of *P* must occupy a part of the same spatio-temporal region that is occupied by the subject which is the bearer of *P*.

As a principle of global ontological significance, SPA has implications that go well beyond the HEC. But before we delve deeper into the metaphysics of realization, it will be helpful to recapitulate the inferential connections that hold between SPA and the HEC. In principle, it seems that one could consistently hold any of the following three positions. (1) If one accepts SPA while endorsing psychological internalism, one ought to reject the HEC. This is Rupert's (2009) orthodox "narrow subjects, narrow properties" view. (2) If one accepts SPA while embracing the HEC, one ought to adopt a "wide subjects" conception of the HEC. This is the position advocated by Clark and Chalmers (1998). (3) Wilson's "narrow subjects, wide properties" view is implied by the combination of accepting the HEC while rejecting SPA.

In Section 4 below, we shall take a critical look at Rupert's "argument from causal interaction" that purports to show that it is metaphysically incoherent to reject SPA. This would eliminate Wilson's option (3) from the menu, which would leave us with only options (1) and (2). But in the next section, I first want to carefully lay out Wilson's reasons for rejecting SPA, and thus further motivate his departure from the "standard view" of physical realization. This means that we have to turn to the metaphysical foundations of functionalism.

6.3.2. Leaving the "standard" view behind

In contemporary philosophy of mind, the notion of *realization* was first introduced with Hilary Putnam's (1960, 1975a) discussion of machine functionalism as a theory about the nature of mental states. Putnam argued that we cannot identify the states of a Turing machine with the physical substrate by which those states are realized, because one and the same machine state can be realized by very different physical substrates. By the same token, mental states cannot be reduced to physical states, because they are multiply realizable in very different physical substrates. Going beyond machine functionalism, non-reductive physicalists have argued that the one-to-many character of the realization relation provides an intuitive conceptual model of how higher-level properties in general are metaphysically dependent on their lower-level realizers yet retain their causal-explanatory autonomy (Fodor, 1974; 1997; Kim, 1993; Horgan, 1993; Melnyk, 2003; Gillet 2002, 2003).

Although the concept of realization has received surprisingly little systematic treatment, the "standard view" of realization can be summarized as a conjunction of two claims (Wilson, 2001, p. 2):

Sufficiency Thesis: Realizers are metaphysically sufficient for the properties they realize.

Constitutivity Thesis: Realizers of properties are exhaustively physically constituted by certain intrinsic, physical states of the individuals (or their parts) whose properties they are.

The "sufficiency thesis" asserts that every higher-level property is metaphysically determined by some cluster of lower-level properties. This is consistent with the multiple realizability of the higher-level property, because it allows that no lower-level realizer is individually necessary for the higher-level property to be instantiated. The "constitutivity thesis" asserts that this metaphysical determination relation is causally grounded in the intrinsic structure of the individual which realizes the higher-level property. The reference to *intrinsic* properties of an individual is meant to exclude properties whose manifestation is metaphysically co-determined by certain relationships between the system and its environment (which makes them *extrinsic* properties). According to Wilson (2004, pp. 22-24), the constitutivity thesis reflects a long-standing metaphysical prejudice in favor of "smallism" – an ontological bias in favor of small things against the not-so-small things and relations which they help to constitute.[72] As applied to the domain of psychology, the constitutivity thesis resonates with the doctrine of psychological internalism, because it implies that the realization of mental states must always be sought in the intrinsic physical properties of individual organisms.

In his writings, Wilson (1995, 2001, 2004, 2005) has extensively argued that the "standard view" of realization provides an inaccurate portrayal of the higher-level taxonomies that are implied by the causal-explanatory practices of the "special sciences" (Fodor, 1974). In biology, for example, there seem to be many kinds of things (e.g., *modules, organisms, species*) that we individuate in terms of various relationships which they bear towards their environment. Moreover, some of the most important properties of those entities (e.g., the *functionality, fitness, adaptedness* of a biological module) are themselves implicitly relational in nature. Contrary to the standard view, the manifestation of those properties does not fully inhere in the individuals that have them. We thus have to look beyond the spatiotemporal boundaries of individuals to determine how these individuals function causally in the world. Reflecting on the causal-explanatory role of such properties, Wilson (2001) has launched a two-pronged attack against the standard view of realization. He argues that the metaphysical context-dependence of their causal roles undermines the conjunction of the sufficiency and constitutivity thesis. To understand the backdrop of Wilson's argument, it will be helpful to review Shoemaker's (1981) influential discussion of functionalism.

According to Shoemaker (1981), a physical state can realize a given functional role only within the causal nexus of the larger system in which it is embedded. Consider, for instance, the property *P* of having a high blood pressure of 140/90. *P* is a property of the mammalian circulatory system which includes the heart, arteries, capillaries, arterioles, venules, blood and other entities. However, not all parts of the circulatory system are equally important for the realization of *P*. In Shoemaker's terminology, the part of the system which is most readily identifiable as playing the most crucial causal role in producing or sustaining *P* is called its *core realizer* (e.g., clogged arteries and a strong heart). This already

---

72 Wilson (2004, pp. 22-24) argues that the meta-physics of "smallism" is a descendant of 17th century "corpuscularian" physics and its explanatory ideal to discover the "intrinsic" properties (viz. primary qualities) of small things. From a methodological perspective, adopting a "smallist" perspective carries an implicit reductionist commitment. "Smallism" implies that we ought to constrain the taxonomies of the special sciences in ways that allow us to *identify* higher-level properties with those entities that the ultimate science of "small things," i.e. physics, discovers.

implies two ways in which the realization-relation is context-sensitive: first, because core realizations are metaphysically sufficient for *P* only when they are taken together with non-core parts of the embedding system; second, because picking out a single causally relevant factor as *the* core realization of P is always dependent on the explanatory context of our functional analysis. In addition, even the *total realizer*, which (for Shoemaker) consists of core plus non-core realizers combined, suffices for *P* only if the right background conditions (e.g., oxygen, gravity...) prevail.

In order to drive a wedge between the sufficiency and the constitutivity thesis, and thus to bring out an inherent tension in the standard view of realization, Wilson (2001) introduces a metaphysical taxonomy of several different species of realizations (Figure 18).

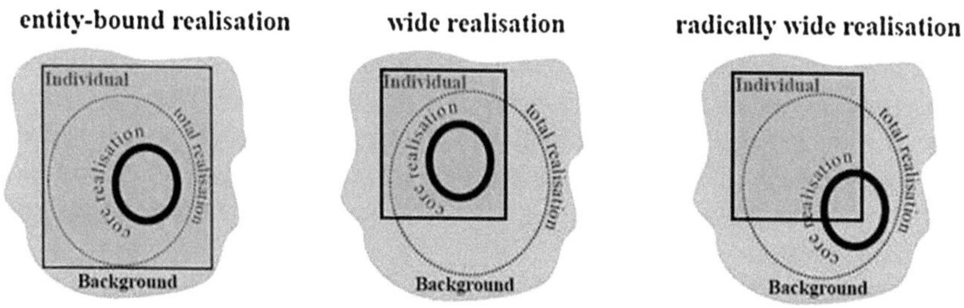

Figure 18 (based on Wilson, 2001). Three species of physical realization.[73]

First, *entity-bound realizations* of a higher-level property *P* are total realizations whose core and non-core parts are completely located within the individual bearer of *P*. Properties with entity-bound realizations are individualistically specifiable in the sense that they pass the "twin test" (Putnam, 1975b). Following Putnam, the twin test involves the (hypothetical) comparison of two individuals that are molecule-by-molecule duplicates, and thus share all the relevant intrinsic physical properties. If some property *P* has an entity-bound realization, the standard view correctly implies that such exact physical duplicates must also share *P*.

But not all properties are like that. According to Wilson, many higher-level properties have realizations whose non-core part is not entirely located within the bounds of the individual; those are properties with so-called *wide realizations*. The manifestation of a widely realized property by its individual bearer violates the constitutivity thesis, since it implies that a certain relationship be-

73 Thanks to Peter Sommerfeld for drawing these diagrams.

tween the individual and its environment is directly implied by the way in which the widely realized property is individuated. Properties with wide realizations fail the "twin test," which shows that the entity-bound portion of their realization is not metaphysically sufficient for their instantiation. If we follow Putnam's (1975b) and Burge's (1979) externalist intuitions about the semantics of propositional attitudes, then intentional mental states like beliefs that *p* or desires that *p* are examples of properties with wide realizations. And more importantly, once we look further into the behavioral, biological, and social sciences, we see that this is by no means an anomaly of folk psychology.

Consider, for instance, the biological property *P* of *being a predator* (Wilson, 2004, pp. 114f). For an animal to be a predator it has to occupy a certain niche within an ecosystem that includes predators and their prey, the dynamics of their causal interactions (e.g., as described by the Lotka-Volterra equations), plus a relevant set of environmental background conditions (e.g., seasonal, terrain conditions). The total realization of *P*, which is the sum of its core realization (e.g., having sharp teeth and claws) and its non-core parts (e.g., the overall speed of the preying animal, the body size and type of the prey), extends beyond the organism into the environment. For instance, if we changed the size or distribution of the prey population, we would thereby affect the nature of the animal as a predator.

Another example of a property with a wide realization is the evolutionary *fitness* of an organism, viewed as its propensity to survive and reproduce in a given environment (Wilson, 2001, p. 13).[74] Many factors which determine an animal's fitness, such as its susceptibility to disease, its vulnerability to predators, and its access to resources are located outside the individual organism. It is metaphysically possible for two organisms to share the exact same physical constitution, but vary in their inclusive fitness in ways that depends solely on their environment. Hence the inclusive fitness of an organism is not wholly metaphysically determined by its intrinsic physical properties, contrary to what the constitutivity thesis asserts.

Wilson also distinguishes a third species of realizations that he calls *radically wide realizations*. These are wide realizations of some property *P* for which not just their non-core part, but also their core part is partly located outside the boundaries of the individual which is the bearer of *P*. For Wilson, social actions such as *making a withdrawal from a bank, being convicted a felon,* or *becoming a homeowner* are among the most clear examples of properties with radically

---

74 The same case can be made for fitness considered as a property of phenotypic traits, as well as other properties of traits such as being an adaptation, being highly specialized, or being a spandrel.

wide realizations.[75] In order to perform the social action of withdrawing money from a bank, the bodily act of signing a certain document is a core part of its realization, but the required type of action can only be performed if the individual stands in appropriate relationships to a vast network of social and institutional conditions (e.g., the financial system, the criminal justice system, and the economic system in the examples above).

According to Wilson, the inherent metaphysical context-sensitivity of certain higher-level properties reveals a dilemma for the standard ("narrow") view of realization. If we decide to hold on to the sufficiency thesis, we ought to include environmental factors, as well as organism-environment interactions, as part of the total realization for properties with a wide (or radically wide) realization. But that would clearly violate the constitutivity thesis. Hence in order to salvage the constitutivity thesis, one would have to give up the sufficiency thesis. But the sufficiency thesis is a central part of a widely shared minimal physicalist consensus (Jackson, 1998). Hence giving up the sufficiency thesis would require a considerable overhaul of physicalism. For most physicalists, admitting the existence of physicalistically kosher properties with "wide" realizations would seem to bear a much healthier ontological cost-benefit ratio. However, properties with wide realizations are incompatible with SPA, the supposedly "general metaphysical principle" (Rupert, 2006, p. 11). But once we reject SPA, Rupert's attempt to generalize his integration argument from cognitive *subjects* to the realization of cognitive *properties* is no longer valid.

In his recent book, Rupert (2009) has launched a metaphysical defense of SPA to deal with Wilson's renegade conception of the HEC (ibid., pp. 78-82). As an example, consider once again the evolutionary fitness of an organism. According to Wilson, fitness qualifies as a widely realized property, because its manifestation is constituted in part by certain relationships between the organism and its environment. In other words, what Wilson calls a "widely realized" property is really an *implicitly relational* property. But the question is: do we really need to admit implicitly relational properties into our ontology, over and above the sum total of intrinsic, non-relational properties and all metaphysically primitive relations? It would seem that an appeal to relational properties is merely shorthand for two distinct states of affairs: first, that an individual has certain intrinsic physical properties, which endow the individual with various dispositions to engage with its environment; second, that an individual stands in an in-

75 For present purposes, I leave it open whether there are interesting cases of (e.g., socially) extended cognition which are best understood as instances of *radically* wide realizations. Establishing this stronger claim is not strictly speaking required for the HEC. See Wilson (2004, 2005a) for a discucssion of "socially manifested" cognitive properties.

definite number of relationships with various things in its environment. Different environmental contexts afford the individual different opportunities for causal interactions. If we change the things with which an individual interacts, this predictably changes the relationships in which it stands. But no further appeal to "relational properties" seems necessary to give a metaphysically complete account of what's going on in this situation.

I have two responses to this die-hard brand of metaphysical reductionism. First, I want to underscore some central features of our causal-explanatory practice of refering to implicitly relational properties (cf. Wilson, 2001, p. 13). To begin with, we should note that if we ascribe a determinate amount of fitness to a biological organism with respect to its environment, we attribute to the organism an implicitly relational *disposition* over and above the sum total of its actual encounters. This disposition is usually described as an individual's ability to survive (i.e., its *viability*) or to reproduce (i.e., its *fertility*) in a certain ecological niche (cf. Sober, 1993, Chapter 3). By assigning a determinate amount of viability to an organism, we measure its disposition to withstand *all relevantly possible causes of death*, i.e., its disposition to withstand not only all the tokens of selection processes that it actually encounters, but also the selection forces which it would encounter in all nearby possible worlds. Without taking into account all the relevantly possible causes of death, attributions of viability fail to express any sensible biological property. This is supported by the "twin-test:" two organisms can differ in their viability solely as a result of changing the set of possible worlds (i.e., environments) which constitute the relevantly possible selection forces. Again, this is not to deny that the intrinsic physical make-up of an organism constitutes an important aspect of the core realization of an organism's fitness. However, it ain't the whole story.

Jennifer McKitrick (2003) has shown that attributions of extrinsic ("wide") dispositions to individuals are common in everyday as well as scientific taxonomies. For instance, we regularly ascribe to objects the power to weigh a certain amount, to open a door, to dissolve a certain substance, to be visible and recognizable. Chemists refer to the *acidity*, *miscibility*, *solubility* of objects; engineers refer to the *conductivity*, *heat sensitivity*, and *rigidity* of objects; and attributions of *vulnerability*, *trustworthiness*, *fertility*, and *stability* are a staple in the social and psychological sciences. The empirical success of scientific generalizations that are couched in terms of such taxonomies provides tentative evidence against the constitutivity thesis. At the very least, it shifts the burden of proof to the metaphysical reductionist, for showing why we should not take these types of causal explanations at face value.

This leads me to my second response, which rests on a certain conception of how we ought to adjudicate metaphysical questions of existence in the context of our debate. Suppose it is true that from the standpoint of the realizer science (e.g., physics, chemistry), we can always in principle break down the wide realization of a higher-level property (and/or disposition) into an individual-bound portion of intrinsic physical properties, together with a set of spatiotemporal and causal relations between the individual and its environment (Prior, Pargetter, & Jackson, 1982). This would imply that implicitly relational higher-level properties (like *fitness*, *trustworthiness*, and *malfunction*) are not part of the most fundamental inventory of our universe. But not many things are, for that matter. So why should that be a decisive move in a debate which is precisely concerned with the ontological commitments of the special sciences, in particular of psychology and cognitive science?

In Chapter 1.5, we have vowed to adhere to the principles of CANDOR, FAIRNESS, and EXTENDED FAIRNESS. CANDOR implies that if properties with wide realizations figure in perfectly intelligible, and empirically supported causal explanations, we should not discriminate against them on grounds of controversial "smallist" assumptions about how scientific taxonomies *ought* to be constrained. Similarly, (EXTENDED) FAIRNESS implies that we should bracket metaphysical qualms about relational properties in the cognitive domain if we regularly do the same in other, non-cognitive domains. Since Rupert's austere brand of metaphysical reductionism flouts all of them, it does not give us an independently plausible reason to doubt the metaphysical coherence of Wilson's "narrow subjects, wide realizations" conception of the HEC. However, Rupert has another argument against Wilson's view up his sleeve, which directly targets Wilson's departure from SPA.

6.4. The argument from causal interaction

To reel in Wilson's renegade metaphysics of wide realizations, Rupert (2006, pp. 12-15; 2009, Chapter 4.4) has presented another, more systematic argument in defense of SPA. It proceeds by way of a good-old-fashioned conceptual analysis of the notion of physical realization.[76] The main point of Rupert's criticism is that mental properties with wide realizations would incoherently imply that certain parts of an organism's environment have to perform a metaphysically dubious kind of double-duty. On the one hand, they would act as external causes and effects of mental states. But at the same time, they would also be constitutive parts of the mental states which they help to cause. Consequently, allowing that properties can have wide realizations would imply that some things (states,

76 My response to Rupert's argument from causal interaction is greatly indebted to the points made by Wilson (2006).

events) are both *causes* of X as well as constitutive *parts* of X. But surely that is impossible, since nothing can be a cause of itself.[77]

As a way of framing his argument, Rupert first proposes a general metaphysical contraint on the relationship between higher-level properties and their lower-level realizers which he calls the "causal-structure principle" (CSP):

Causal-structure principle (CSP): The token realizer of a property M must enter into a network of causal relations a subset of which is isomorphic (in actual as well as in certain counterfactual cases) to the network of causal relations into which M enters (Rupert, 2009, p. 66).

For instance, if a cognitive state-token *P* causes (or is caused by) state-token *Q*, then CSP implies that the realizer of *P* must cause (or be caused by) a realizer of *Q*. Rupert apparently thinks that the CSP follows from the sufficiency thesis of the "standard view" of realization (cf. Chapter 6.3 above).[78] Spelled out in full detail, Rupert's argument from causal interaction runs as follows (Rupert, 2009, pp. 69f):

1. According to functionalism, the causal profile of a cognitive state M of an organism O is partly determined by O's causal interactions with certain objects of O's external environment E.
2. According to the HEC, the physical realization P_E of states of E (e.g., Otto's notebook) is an extra-organismic part of the total physical realization P_T of M (e.g., Otto's extended belief).
3. P_E does not causally interact with P_T.

Therefore, P_T's causal profile does not contain M's causal profile as a subset.

4. According to CSP, property X realizes property Y only if X's causal profile contains Y's as a subset.

Therefore, P_T is not a realizer of M.

Since the conclusion would hold of higher-level properties in general, it would entail, among other things, that Wilson's "narrow-subjects, wide proper-

77 There is a widely albeit not universally shared consensus that the notion of self-causation is metaphysically incoherent. I take it that HEC-theorists do not wish to be committed to a highly controversial metaphysical doctrine, especially since a much less contentious interpretation of Wilson's view is right around the corner, as I shall argue.

78 Rupert (2009, p. 68) imposes several further methodological constraints on CSP that are not directly implied by the sufficiency thesis. However, since the crux of his argument does not depend on any of those contraints, we can side-step the question whether they are independently plausible.

ties" conception of the HEC is metaphysically incoherent. But is Rupert's argument sound?

6.4.1. Parts, wholes, and their causes

Let us now carefully attend to Rupert's argument from causal interaction. Consider Premise (3), which reflects the assumption that no part of a realization of X can also be a cause of X. As we have noted, the reason behind Premise (3) is a metaphysical injunction against a potentially incoherent form of self-causation. However, the same reasoning does not rule out that some entity Z cannot at the same time be both a constitutive part of X and also a cause of Y. In particular, this remains true in situations *where Y itself is a proper part of X*. Let me describe some situation in which this seems to be the case.

To begin with, let X be an ordinary mental property with an organism-bound realization. For the sake of the argument, let us consider pain as mental property whose total realization is entirely located within the bounds of the organism that experiences it. A (radically simplified) functionalist theory of a pain-type mental state might go like this: Pain is a functional state M that is caused by pin pricks and which causes worry and the emission of loud noises, and worry causes brow wrinkling (Block, 1996). Consistent with the multiple realizability of mental states, the functional characterization of pain states does not tell us anything specific about the causal mechanism which is responsible for implementing the pain-role (e.g.) in humans. But if we look for such a mechanism, we would expect to find a chain of causal connections among different, non-overlapping parts of the complex organic state which is the total realization of pain in human beings.

For example, consider the overall state of the nociceptive system N as an individual-bound total realization of M in humans. N begins with the mechanical and polymodal receptors in the skin, runs through myelinated and unmyelinated axons, the spinal neurons, parts of the brainstem and the cerebellum, and eventuates in the somato-sensory area of the cortex. Even though these physical structures are all part of the total realization of the pain-state, there is nothing mysterious about the discovery of a causal chain that connects these different parts of the nociceptive system. For instance, the activation of a mechanical receptor in our leg, caused by some external stimulus like a pin prick, causes our C-fibres to fire, which in turn causes a state change further downstream in the somatosensory areas of the cortex. Thus far, nothing in our story should raise any metaphysical eyebrows.

So what's supposed to be so problematic about functional properties with wide realizations? Here, too, we would expect that if X is the wide realization of a mental property M, the *external* portions of X can causally interact with non-overlapping *internal* portions of X. As an example, consider Otto's extended belief M that MoMA is located on 53rd Street. The causal profile of M includes the coarse-grained behavioral disposition to walk towards 53rd Street provided that one also believes that there is a new art exhibition at the MoMA, and has the desire to see it. According to the HEC, Otto and Inga share the same coarse-grained behavioral disposition to walk towards 53rd Street in virtue of sharing M, except that the physical machinery P_T which for Otto is the total realization of M partly extends into the environment (i.e., into Otto's notebook). But as before, this does not preclude any causal interactions among distinct parts of P_T from occurring. For instance, when Otto hears the exciting news, some neurons in his brain begin to fire, which causes him to pull out his notebook, flip through its pages, search for the relevant entry, read the entry, which causes neural firings in his visual cortex, etc. Still, there is nothing metaphysically spooky about any of this. Something must be wrong with Rupert's argument. But what is it?

My diagnosis is that Rupert crucially fails to take into account that *not all causal interactions are equal* for a functionalist theory of mental states (Chapter 2.3). Rupert is right that for the extended functionalist, environmental states play two different roles in a theory of an extended mental state M. On the one hand, they act as causes and effects of M; on the other hand, they act as extended realizers of M. Rupert erroneously assumes that it has to be *one and the same part* of the world which does both things at the same time, but that is not the case. Hence there is no metaphysically dubious double-duty here, only *metaphysically kosher single duties that are performed by distinct parts of the world.* Relative to an extended functionalist theory of M, those parts of the environment which figure among the role-individuating causes and effects of M are not part of the total realization base for M. Conversely, the causal interactions among elements of the total realization base for M do not concern our individuation of the relevant functional role of M.

Now we can see more clearly where Rupert's argument breaks down. It equivocates in its premises between the two distinct roles (E_1 vs. E_2) that some environmental state E can play vis-à-vis an extended mental state M. On the one hand, if E_1 is analyzed as a typical *cause* or *effect* of M, then Premise (1) is true, but Premise (2) is false: because the HEC does not consider E_1 as a part of P_T. For instance, Otto's disposition to walk towards 53rd street, contingent on hearing about the new art exhbit at MoMA and the desire to see it, is a typical *effect* of his extended belief that MoMA is on 53rd Street. But Otto's walking behavior is not considered part of the physical realization of M. On the other hand, if E_2 is analyzed as a *part* of P_T, then Premise (2) is true, but Premise (1) is false: because the HEC does not consider the causal interaction with E_2 as a feature of M's typical causal-functional role. For instance, the movements of Otto's fingers as he flips through his notebook are *part* of the physical realization of M. However, they are not considered as an essential feature of the causal-functional role of M. Once we properly distinguish between these two conceptually distinct roles of environmental resources, the alleged tension within extended functionalism quickly disappears.

The flaw in Rupert's argument from causal interaction does not hinge on any idiosyncratic features of the cognitive domain, either, but can be demonstrated with respect to other, non-cognitive properties. Once again, consider *adaptedness* as a widely realized biological property of organismic traits.[79] As a concrete example, consider how the flight response of the antelope is well-adapted to its species-typical environment. Antelopes base their decision to flee on two factors: the type of predator which they encounter, and the predator's relative distance to the herd. For instance, an antelope permits lions that are fully visible to come within 200 meters before fleeing, because the fact that a hunting lion prefers to hide while stalking its prey makes it unlikely that a visible lion will attack. But antelopes flee from cheetahs when they are over 800 meters away, because cheetahs are known to attack by sprinting.[80]

Following Wilson, we can say that the adaptedness of the antelope's decision strategy is partly *realized* by all the relevantly possible causal interactions between antelopes and their potential predators in a species-typical environment, because it constitutively depends on such factors as the high speed and sharp teeth of the cheetah. Rupert is right to insist that it would be incoherent to say that (e.g.) the sharp teeth of the cheetah *cause* the adaptedness of the antelope's decision strategy. However, it is perfectly coherent to say that the speed of the cheetah is causally responsible for catching the antelope, or that its sharp teeth are causally responsible for biting deep into the throat of its prey, etc. – a causal chain of non-overlapping events which, if they were to occur, would most likely kill the antelope.

6.4.2. The piecemeal causation of total realizers by their parts

To put the final nail into the coffin of Rupert's argument from causal interaction, let us return to Premise (3) once more. It is based on the assumption that no part of a physical realization of X can also be a cause of X. We provisionally ac-

---

79 I use the notion of *adaptedness* in a purely non-historical sense, as referring to the current utility or adaptive value of an organismic trait with respect to a certain environment. The adaptedness of a trait should thus be distinguished from the causal-historical notion of an *adaptation*, which must be the result of natural selection.

80 See the Wikipedia entry: http://en.wikipedia.org/wiki/Antelope.

cepted Premise (3), but aren't there situations in which wholes and their parts appear to be causally connected? A well-known example from action theory will help to illustrate this possibility.

Consider John Wilkes Booth's assassination of Abraham Lincoln on Good Friday, April 14, 1865 during a play at Ford's Theatre. Around 10 p.m., after seeing through his spyhole that Lincoln had indeed made it to the play, something went on in Booth's brain. Perhaps a few neurons in his prefrontal cortex were firing; the neural signals travelled through his brain into the body. Then his muscles contracted – and as they did, he bent his finger; and by doing so, he pulled the trigger of his .44 caliber Deringer. A shot was fired which eventually killed Lincoln. How many actions did Booth perform that day when he killed Lincoln? Did he first pull the trigger, and then shoot Lincoln? Most of us would say that Booth did not have to perform two (or more) distinct actions when he shot Lincoln, but a single action that can be *described* in different ways (cf. Anscombe, 1957, pp. 45f). Now suppose further that we follow Davidson's (1963, 1985) analysis of action sentences as quantifications over events (that can be described in multiple ways). Then the following puzzle arises (Pietroski, 2000, Chapter 1). Booth shot Lincoln with a pistol, but he did not pull the trigger with a pistol. Booth pulled the trigger with his finger, but he did not shoot Lincoln with his finger. Given the principle of the indiscernibility of identicals, doesn't the discernibility of those two descriptions mean that we are talking about two distinct events? Hence assuming that we continue to accept Davidson's analysis, didn't Booth have to perform multiple actions, after all, when he shot Lincoln?

One way to solve this puzzle is to say that Booth performed one basic action, and in virtue of performing it, he caused several distinct but causally related events (Pietroski, 2000, Chapter 1). On a counterfactual analysis of causation, conjoining the following two conditionals adds up to the whole action of *shooting Lincoln*, i.e., a causal chain of events that originates in a change in Booth's brain and terminates in Lincoln's death:

1. If Booth would not have bent his finger, he would not have pulled the trigger.
2. If Booth would not have pulled the trigger, he would not have shot Lincoln.

However, there still remains a problem for this analysis. Since the event p of Booth bending his finger is a proper part of the shooting event q, p and q are not wholly distinct. This creates the following dilemma (Lewis, 1986, pp. 172-75). If we allow that an event q can be causally dependent on another event p from which it is not wholly distinct (e.g., because it is among its parts), this would imply that q is causally dependent on itself, contrary to what we have assumed. If, on the other hand, we entirely rule out part-to-whole causation, then we can no longer say that q is causally dependent on p. In that case, we'd still have no solution to the problem of multiplying the number of actions which Booth had to perform that day in order to kill Lincoln.

As a way out of this dilemma, Lewis (ibid.) has suggested to break down the complex event q of shooting Lincoln into a sequence of smaller events. Those smaller events can be seen as forming a series of causal interactions which occur only between non-overlapping parts of the complex q-event. For instance, we can say that an early part of the complex shooting event (e.g., the organism-internal parts of q) causes a later part of the same complex shooting event (e.g., the organism-external parts of q) to occur. Based on the availability of the suggested decomposition, Lewis then defines the notion of a "piecemeal cause" as a weaker substitute for part-whole "causation:"

PIECEMEAL CAUSATION: c is a *piecemeal cause* of e if and only if c and e are divisible into parts in such a way that every part of e is causally dependent on (directly or connected by a causal chain) some part of c (ibid., p. 172).

The distinction between causation and piecemeal causation can be used to further elucidate the metaphysics of wide realizations. We have seen earlier how the external portion P_E of an extended realizer P_T (instantiated at t_1) can be a *cause* of a different, non-overlapping internal portion of P_T (instantiated at time t_2). Now we can also see how the external portion P_E of an extended realizer P_T (instantiated at t_1) can be a *piecemeal cause* of P_T (instantiated in its entirety only at time t_2). For instance, the external portion of Otto's extended belief M at time t_1 (i.e., the inscription in his notebook before he accesses it) is a piecemeal cause of the more encompassing total realization P_T of M at time t_2, which includes the brain-bound portion of P_T.

Rupert (2009, p. 73) seems to anticipate a rejoinder along the lines that we have just sketched.[81] He offers two reasons to be skeptical of the "deviant" (ibid.) manner in which the exernal portion P_E at t_1 could be said to "cause" an extended total realizer P_T at t_2.

His first reason is simply that "independently causing distinct proper parts of a complex event is not the same as causing the event *simpliciter*" (ibid.). But this seems to be the correct diagnosis of our solution rather than the statement of a problem. First, it is not clear that a HEC-theorist ought to be committed to an underlying metaphysics of event causation which is significantly stronger than

81 Since his discussion does not explicitly mention Lewis' idea of *piecemeal causation*, it is not clear whether he has any specific proposal in mind.

counterfactual dependence relations (Lewis, 1973, 1986). But more importantly, even if we grant that piecemeal causation is weaker than direct causation, this would not affect the coherence of Wilson's extended functionalism. This is because the derivative relation of piecemeal causation does not crucially play into the definition of causal-functional roles, nor does it directly enter the construal of the realization relation. Hence Rupert's worry that piecemeal causation is not quite as strong as causation proper is a moot issue.

Rupert's second concern is that the interaction between organisms and their environment by way of piecemeal causation is not "intimate" (ibid., p. 73) enough to support the HEC. In effect, he argues that once the external portion of the total realizer of M has done its job causing (at t_1) the internal portion of M (at t_2), and perhaps also piecemeal-causes the entire realizer of M (also at t_2), it drops out of the picture as far as the causal efficacy of M is concerned. So why should we count the external portion as a constitutive part of the "wide" realization of M in the first place?

My answer to this worry has two parts. Concerning the causal-functional role of extended mental states, I have nothing new to add here to the arguments in favor of extended functionalism that we have already discussed in Chapter 2. So let us focus on the problem of causal intimacy, which Rupert mentions in the specific context of Haugeland's (1998) classic argument for the HEC. Understood in this manner, the problem of causal intimacy is arguably more pertinent to Clark & Chalmers' "wide subjects" view than it is for Wilson's own view, although this may ultimately be a matter of degree and explanatory emphasis (see also Chapters 3.4 and 6.2 for discussion). More importantly, though, I don't see why piecemeal-causation is, generally speaking, too weak to support "intimate" forms of causal intercourse.

As a case in point, consider Lewis' (1986, pp. 172-75) own exemplar of a self-sustaining causal process which exhibits piecemeal causation: the howling which occurs in a public address system when the sound emitted from the speakers travels back into the microphone, is re-amplified, and sent through the speakers again. It seems to me that this feedback loop, which results in howling, is in fact an apt metaphor for the mutually reinforcing, co-adaptive spiral between neural, bodily, and environmental processes which we have characterized as extended cognition. For instance, according to the "interactive vision" framework (cf. Ballard, 1996; Ballard et al., 1997; Churchland, Ramachandran, & Sejnowski, 1994; see also Clark, 2008; Wilson, 2010), perception and action continuously feed into each other at all levels of visual processing. Lower-level perceptual processes directly trigger motor routines (e.g., head, neck, and eye movements) whose epistemic role is to generate enhanced sensory input. This in turn enables an improved visual protoanalysis of the perceived scence, which activates further visuomotor actions, etc. In this temporally extended series of perception-action cycles, perceptual and motor processes incrementally cause one another piecemeal: changes in one's perceptual input drive one's motor actions, and changes in one's bodily position yield new self-generated perceptual input. As far as I can see, the metaphysics of piecemeal causation is entirely compatible with an intimate causal commerce among the internal and external portions of extended cognitive processes.

7. Conclusion

Our collective fascination with *liminal beings* – creatures that are seemingly caught between two natures and thus elude conventional systems of classification – is revealed in a large body of cultural folklore and literary fiction (Turner, 1969). Part human and part horse, the centaurs of Greek mythology are a hybrid race which straddles the divide between biological species. Reports about the apparition of ghosts, creatures which straddle the divide between the living and the dead, are a widespread cultural phenomenon. Our information age harbors a predilection for cyborgs ("cybernetic organisms") which span the divide between man and machine. The inherently ambiguous nature of liminal beings makes for an uncanny appearance but at the same time, it can be uniquely instructive. For instance, the centaurs of Greek mythology usually embody the untamed nature and the lower appetities of mankind, as depicted in their battle with the Lapiths; but occasionally, centaurs can also symbolize wisdom and civility, as embodied in the venerable teacher Chiron.

In this book, I have argued that the human mind itself is a liminal entity many of whose distinctive accomplishments straddle the biological boundaries between brain, body, and environment. The "extended mind" thesis flouts conventional wisdom that the cognition is entirely located inside the head – a view that we have dubbed "Cartesian materialism" (Chapter 1.1). For Descartes, it seemed evidently true that minds are essentially non-extended things. Our scientifically enlightened Cartesian materialists pride themselves on having exorcised the Cartesian ghost from the biological machine, but nevertheless remain committed to the Cartesian vision of the mental as something purely *inner*. Against their lopsided emphasis on the brain, the extended mind thesis stresses the importance of dynamic brain-body-environment couplings, the opportunistic exploitation of bodily morphology, the canny performance of epistemically potent bodily actions, the generation and use of external representations, and the multiple layers of cognitive scaffolding provided by tools, aids, and social-cultural practices. My central thesis is that extended minds like ours arise only from the functional integration of a coalescing ensemble of neural, bodily, and environmental resources.

Not everybody agrees that this deeply promiscuous view of the mind provides a sensible alternative to Cartesian materialism as a scientific framework for the studying intelligence and cognition. Clinging to the biological fortress of the brain, hard-nosed Cartesian materialists still scoff at the prospect of a reformed cognitive science in which the notions of embodiment, epistemic action, and cognitive extensions take center stage. They may acknowledge some of the empirical discoveries that have been made under the banner of situated, embodied, and distributed approaches to cognition (Chapter 1.4), yet vigorously resist their more fundamental ontological implications. For the Cartesian materialist, the body remains merely an expedient *instrument* for the mind, whose activities are causally, but not constitutively dependent on *inputs* delivered by the environment. Cognition proper springs from the brain only.

What drives this deep-seated parochialism about the mind? What makes the idea that cognition literally extrudes beyond the brain into the body and the world so difficult to accept? Speaking of centaurs – to take a fictional example – would we be equally inclined to insist that a centaur is *really* a human being, albeit with a physique that "depends heavily, in hitherto unexpected ways" (cf. Rupert, 2004, p. 393) on being conjoined to a horse's body? Why should we care about where exactly man ends and horse begins? For instance, if we had to explain how centaurs fight, we would probably be much more interested in where and how exactly their limbs are conjoined, or what special sorts of qualities they gain from their hybrid assembly of heterogeneous body parts.[82] But Cartesian materialists tolerate no such pragmatism when it comes to the human mind and all things mental. So perhaps the subtle, but lasting legacy of Descartes' mind-body dualism is the Cartesian materialists' assumption that there must be an indubitable "mark of the mental" – a metaphysically primitive, essential property of mental states. Knowing that feature would allow an omniscient observer to divide the world into exactly two kinds of things: mental vs. non-mental things. The essentialist quest for such a demarcation criterion strikes me as the ultimate remnant of Cartesian metaphysics.

But such a quest may well be misguided. Using the famous parable of the blind man with his stick, Gregory Bateson (1972/2000) has nicely illustrated how our ways of individuating cognitive systems and/or processes are inextricably bound up with the intents and purposes of psychological explanation:

> "Suppose I am a blind man, and I use a stick. I go tap, tap, tap. Where do *I* start? Is my mental system bounded at the handle of the stick? Is it bounded by my skin? Does it start halfway up the stick? Does it start at the tip of the stick? But these are nonsense questions. The stick is a pathway along which transforms of difference are

---

82 Historically speaking, there was a considerable flux in the artistic representation of centaurs (Roscher, 1894, pp. 1074-79). In early Attic and Boeotian vase-paintings, for example, centaurs are depicted as humans with the hindquarters of a horse attached to them – a symbolization which subordinates the animal body to the human body. But in later renderings, which became the culturally dominant standard, centaurs have the torso of a human that is joined at the waist to the horse's withers, i.e., by replacing the parts from the horse's neck upwards. This classic mode of representation evidently places a greater emphasis on the supposedly brute, animalistic nature of centaurs.

> being transmitted. The way to delineate the system is to draw the limiting line in such a way that you do not cut any of these pathways in ways which leave things inexplicable. If what you are trying to explain is a given piece of behavior, such as the locomotion of the blind man, then, for this purpose, you will need the street, the stick, the man; the street, the stick, and so on, round and round. But when the blind man sits down to eat his lunch, his stick and its messages will no longer be relevant – if it is his eating that you want to understand" (ibid, p. 465).

Following Bateson's lead, the best way to defeat Cartesian materialism is to demonstrate that a strictly brain-bound (or even organism-bound) psychology sometimes cuts mental pathways in ways which *does* leave certain behaviors unexplained, or at least not as well explained as by the extended mind thesis. Hence my defense of the extended mind thesis has been predicated on showing two things: first, that psychologists sometimes profitably treat neural, bodily, and bio-external resources as parts of a single, hybrid *system* (and/or *process*); second, that we have good reasons to consider such systems (and/or processes) as *cognitive*. Let me briefly recapitulate some of my central arguments in support of the hypothesis of extended cognition (HEC).

First, what do we mean by *cognition*? Because of the sheer heterogeneity of psychological predicates that one may want to include in the category of mental or cognitive properties, I have advocated a "big-tent" approach (Chapter 1.4) that includes traditional folk-psychological predicates (e.g., belief, desire, intention), predicates for particular cognitive capacities (e.g., memory, reasoning, problem-solving, motor control), and more abstract construals of cognition that have proven to have explanatory value in cognitive science and psychology (e.g., intentionality, representation, computation, behavioral control, coordination dynamics). It makes good sense to call all these properties "cognitive," based on a firm but defeasible pre-theoretical understanding of the distinctive roles which they play in the production of intelligent behavior (cf. the principle of parity, Chapter 2). But contrary to a widely shared presupposition in the current debate, our big-tent notion of the cognitive does not refer to, or even point to some underlying essential features of cognition. Instead, it mostly acts as a placeholder for particular psychological predicates that are used to describe concrete psychological models. Hence we could in principle dispense with the philosophically loaded term "cognition," and replace it with more specific terms such as "belief," "memory," or "reasoning." Rather than arguing for the "hypothesis of extended cognition," one really has to argue separately for the "hypothesis of extended beliefs," the "hypothesis of extended memory," or the "hypothesis of extended reasoning." Keeping that caveat in mind, however, I see no harm in our expository practice of generically referring to various states, properties, capacities, and tasks as "cognitive," assuming that they bear important family resemblances to each other.

Since we currently have multiple theoretical frameworks for studying cognition in its many facets, several routes can be taken to support the HEC. Based on three meta-philosophical ground rules for our debate (Chapter 1.5), I have presented several arguments all of which are firmly rooted in the explanatory practices of contemporary cognitive science, and illustrated each route with detailed case studies (Chapters 3-5). Let me summarize two of the main arguments in my book.

One route is to consider the organization of cognitive and sensorimotor processes which arise from the complex, densely coupled, and often highly nonlinear interactions between brain, body, and environment (Chapter 3). In Chapter 3, I reflect on the use of dynamical systems theory to study cognitive and sensorimotor processes as they unfold in time. In particular, I discuss the multilayered structure of Beer's dynamicist model of *minimally cognitive behavior*, and draw attention to Bingham's analysis of the organization of human action in terms of the dynamics of *task-specific devices*. Based on my reconstruction of the "dynamicist stance" in cognitive science, I argue that the successful practice of dynamicist modeling supports the following claims. (1) The best explanation of (at least certain types or aspects of) intelligent behavior is likely to involve the dynamical analysis of (esp. nonlinearly) coupled brain-body-environment systems. (2) Dynamicist analyses have a distinctively "top-down" style of explanation which is epistemically irreducible to the traditional "decomposition-and-localization" style of explanation. (3) What makes a dynamical system cognitive is its capacity to generate particular types of intelligent behavior. Taken together, these largely epistemological considerations suggest an inference to the HEC based on our best (dynamicist) explanations of intelligent behavior.

In Chapter 5, I develop in great detail a representational argument for the HEC. When mental representations are posited as theoretical entities in psychological explanations of behavior, they are expected to do double-duty: first, they ought to have the right kind of *representational* content; second, they ought to play a relevant *causal* role in the production of intelligent behavior. These two requirements constrain the kinds of states which are apt to play the role of mental representations. As part of my argument for the HEC, I carefully describe a situation which requires that we establish the cognitive type-identity of two different token-processes of syllogistic reasoning, both of which involve the manipulation of external representation systems. In one case, a person uses the diagrammatic method of Euler Circles; in the other case, a different person uses the sentential method of Natural Deduction. The challenge I raise is to say on what grounds we are entitled to consider both token-processes as episodes of *syllogis-*

tic reasoning. In response to my challenge, I propose a "division of cognitive labor" hypothesis. On the one hand, external representations are dependent on perceptually grounded representational mechanisms inside the brain to have a causal influence on our behavior. On the other hand, the internal representations on which the neural mechanism operates are dependent on our use of external representations to have their syllogistic content fixed. Thus if we individuate cognitive processes in terms of the distinctive, non-derived content of the representations on which they operate, it is only the functional integration of internal and external representational vehicles which qualifies the extended computational processes as two bona fide episodes of *syllogistic reasoning*. Therefore, we ought to embrace the HEC, rather than its conservative internalist rival, the hypothesis of embedded cognition (HEMC).

Computationalism and dynamicism are often seen as mutually incompatible approaches to the study of mind and cognition (van Gelder, 1995; Port & van Gelder, 1995; Spivey, 2007; Chemero, 2009). Whether this is really true remains a subject of considerable debate that I won't try to resolve here (for an ecumenical approach, see Clark, 1997; 2008). More significant though, at least from our ecumenical perspective, is the discovery that two theoretically independent approaches provide converging evidence against the doctrine of Cartesian materialism, and in favor of our hybrid vision of the extended mind.

Finally, there is one last issue we ought to address. A well-disposed reader may still have the lingering suspicion that this debate is ultimately a verbal dispute over how we choose to apply the label "cognitive." In particular, the bodily and environmental processes that may pass muster as instances of extended "cognition" are not the real (cognitive) McCoy. They are "cognitive" insofar as they are *metaphorical* extensions of vocabularies and theories that are *literally* true only of certain neural processes in the brain. How should we respond to such a criticism?

To be sure, there is nothing disreputable about the use of analogies in science. It has been shown that analogies promote the discovery, development, evaluation, and exposition of scientific ideas (Holyoak & Thagard, 1995). But there is a deeper reason for the intuitive pull of the *principle of parity* (Chapter 2) between neural and extra-neural resources which lies buried deep in the history of the very conceptual frameworks on which cognitive science was built. This sometimes forgotten aspect of its history was brought back into light in a lucid discussion by Edwin Hutchins (1995a, Chapter 9).

Consider, for instance, the computational theory of mind. Is it a "metaphorical extension" to consider the nautical charts on which a navigator plots the course of his ship as part of his computational apparatus? Is it merely an "analogy" to say that the ancient accountants of the Middle East used their clay tokens and pictographic tablets for their calculations? Surely this would be to confuse the base domain from where the notion of *computation* originates with the target domain (i.e., the brain) to which it can be – sometimes profitably – applied. As Hutchins (1995a, p. 364) summarizes his alternative history of cognitive science,

> "The computer was made in the image of the sociocultural system, and the human was remade in the image of the computer, so the language we use for mental events is the language that we should have used for these kinds of sociocultural systems to begin with."

Hutchins' account is particularly compelling as a reconstruction of classical symbol-based computational models of human intelligence, such as Newell and Simon's (1976) "physical symbol systems hypothesis." From a historical perspective, the pre-history of symbolic computation clearly begins outside the head, with the cultural evolution of writing systems and specialized notations which eventually led to the development of formal systems. A significant point of this development was the invention of machines that could automatically manipulate formal systems: the digital computer. Now we finally had machines that could perform difficult computational tasks that would require intelligence if they were done by a person (Turing, 1950). This accomplishment is undoubtedly what makes the computer such an impressive feat of "forward engineering."

It is therefore quite natural that cognitive psychologists subsequently began to draw on our understanding of computers – man-made machines that we evidently knew how to build – in order to explain how the mind works. The idea to "reverse-engineer" the mind/brain, i.e., to explain its functioning by way of a computational analysis of its design, has been a hallmark feature of orthodox cognitive science (Newell & Simon, 1976; Dennett, 1978; Marr, 1982). But when cognitive scientists took the computer as a model of the mind/brain, they forgot that the "embodied" activities of a person who carries out computations by hand, and with pen and paper, had earlier dropped out of the picture as irrelevant for the purpose of forward-engineering automatic formal symbol manipulators. Hence the "physical symbol-system architecture" which became the dominant cognitive-scientific model of the mind/brain was in fact already an abstract, idealized model of computation from which the adroit symbol-manipulator – the human person with hands and eyes – had already been removed (Hutchins, 1995a, p. 363). A failure to notice the "conceptual surgery" (ibid.) which preceded the application of the computer model to the biological mind/brain meant that classical cognitive science was bound to view human cognition as an essentially disembodied phenomenon – as "*rational,* in that deliberative, conceptual thought is viewed as the primary exemplar of cognition; *abstract,* in the sense that implementation and the nature of the physical environment are treated as of secondary importance [...]; *detached,* in the sense that thinking is treated separately from perception and action" (Smith, 1999, p. 769; see Chapter 1.2.1). Later, once the reverse-engineering project was in full swing, it must have seemed intuitive to the cognitive scientist to identify the boundaries of the computational mind with the biological boundaries of the brain. Finally there was scientific "proof" of Cartesian materialism!

Hence it would be a colossal symptom of collective memory loss if we were to regard the extended mind thesis merely as a "metaphorical extension" of psychological theories that are "literally" true only of the brain. *Au contraire*: cognitive scientists had tacitly applied a principle of parity when they shifted their unit of analysis from individuals in their socio-cultural environment to the brains of those individuals, long before Clark and Chalmers (1998) suggested that we finally ought to move back again from the brain into the world. Proponents of the HEC are not guilty of committing a "coupling-constitution" fallacy (Adams and Aizawa, 2008) if they look beyond the brain for the mechanistic basis of our cognitive abilities. By mistaking the cognitive properties of embodied, situated, and culturally embedded systems for the intrinsic properties of individual brains, Cartesian materialists were guilty of committing a "subtraction" fallacy. It is now incumbent on proponents of the HEC to set our sights aright.

The time has come to sum it all up. At the outset of our book (Chapter 1.1), I hailed the extended mind thesis as the latest attack on our narcissistic human pride, i.e., the Cartesian materialists' belief that our minds are fully contained inside the brain. Should we thus conceive of psychological externalism as yet another "mortification of humanity," as Freud would have put it? Based on our discussion, I would argue that the real mortification of the human mind would be to lock it up inside the brain, and thus cut it off from the cognitive fuel provivded by our bodily, gestural, material, technological, and socio-cultural scaffolds. Hence externalism is a reason for joy – not grief – because it reminds us that the ultimate self-empowerment of the mind comes from extending itself into the world.

8. References

Abraham, R., & Shaw, C. (1992). *Dynamics: The geometry of behavior*. Redwood City, CA: Addison-Wesley.

Ackerman, M. S., & Halverson, C. A. (2000). Re-examining organizational memory. *Communications of the ACM, 43*, 58-64.

Adams, F., & Aizawa, K. (2001). The bounds of cognition. *Philosophical Psychology, 14*, 43-64.

Adams, F., & Aizawa, K. (2008). *The bounds of cognition*. Oxford: Blackwell.

Adams, F., & Aizawa, K. (2009). Why the mind is still in the head. In P. Robbins & M. Aydede (Eds.), *The Cambridge Handbook of Situated Cognition* (pp. 78-95). New York: Cambridge University Press.

Anderson, J. (2005). Human symbol manipulation within an integrated cognitive architecture. *Cognitive Science, 29*, 313-341.

Anderson, M. L. (2003). Embodied cognition: A field guide. *Artificial Intelligence, 149*, 91-130.

Anscombe, G. E. M. (1957). *Intention*. Oxford: Blackwell.

Antelope. Retrieved June 7, 2010, from Wikipedia, the free encylopedia: http://en.wikipedia.org/wiki/Antelope

Argote, L. (1999). *Organizational learning: Creating, retaining and transferring knowledge*. Norwell, MA: Kluwer.

Baeza-Yates, R., & Ribeiro, B. (1999). *Modern information retrieval*. Harlow, UK: ACM Press.

Balasubramaniam, R., & Turvey, M. (2004). Coordination modes in the multisegmental dynamics of hula hooping. *Biological Cybernetics, 19*, 176-190.

Ball, P. (1998). *The self-made tapestry: Pattern formation in nature*. Oxford: Oxford University Press.

Ballard, D., Hayhoe, M., Pook, P., & Rao, R. (1997). Deictic codes for the embodiment of cognition. *Behavioral and Brain Sciences, 20*, 723-767.

Ballard, D. (1996). On the function of visual representation. In K. Akins (Ed.), *Perception* (pp. 111-131). New York: Oxford University Press.

Baltes, P. B., Reuter-Lorenz, P., & Rösler, F. (2006). *Lifespan development and the brain: The perspective of biocultural co-constructivism*. New York: Cambridge University Press.

Barandiaran, X., & Moreno, A. (2006). On what makes certain dynamical systems cognitive. A minimally cognitive organization program. *Journal of Adaptive Behavior, 14*, 171-185.

Barkow, J. H., Cosmides, L., & Tooby, J. (1992). *The adapted mind: Evolutionary psychology and the generation of culture*. New York: Oxford University Press.

Baron-Cohen, S. (1995). *Mindblindness: An essay on autism and theory of mind.* Cambridge, MA: MIT Press.

Barsalou, L. W. (1999). Perceptual symbol systems. *Behavioral and Brain Sciences, 22*, 577-609.

Bateson, G. (2000). *Steps to an ecology of mind: Collected essays in anthropology, psychiatry, evolution, and epistemology*. Chicago: University of Chicago Press.

Beach, K. (1988). The role of external mnemonic symbols in acquiring an occupation. In M. M. Gruneberg, P. E. Morris & R. N. Sykes (Eds.), *Practical aspects of memory* (pp. 342-346). Chichester, UK: Wiley.

Bechtel, W. (1998). Representations and cognitive explanations: Assessing the dynamicist's challenge in cognitive science. *Cognitive Science, 22*, 295-318.

Bechtel, W. (2008). *Mental mechanisms: Philosophical perspectives on cognitive neuroscience*. London: Routledge.

Bechtel, W., & Richardson, R. C. (1993). *Discovering complexity: Decomposition and localization as strategies in scientific research*. Princeton, NJ: Princeton University Press.

Beer, R. D. (1995). A dynamical systems perspective on agent-environment interaction. *Artificial Intelligence, 72*, 173-215.

Beer, R. D. (1996). Toward the evolution of dynamical neural networks for minimally cognitive behavior. In P. Maes, M. Mataric, J. A. Meyer, J. Pollack & S. Wilson (Eds.), *From animals to animats 4: Proceedings of the Fourth International Conference on Simulation of Adaptive Behavior* (pp. 421-429). Cambridge, MA: MIT Press.

Beer, R. D. (2000). Dynamical approaches to cognitive science. *Trends in Cognitive Sciences, 4*, 91-99.

Beer, R. D. (2003). The dynamics of active categorical perception in an evolved model agent. *Adaptive Behavior, 11*, 209-243.

Beer, R. D. (in press). Dynamical systems and embedded cognition. In K. Frankish & W. Ramsey (Eds.), *The Cambridge Handbook of Artificial Intelligence*. Cambridge, UK: Cambridge University Press.

Beer, R. D., Chiel, H. J., & Gallagher, J. C. (1999). Evolution and analysis of model CPGs for walking II. General principles and individual variability. *Journal of Computational Neuroscience 7*, 119-147.

Bingham, G. P. (1988). Task-specific devices and the perceptual bottleneck. *Human Movement Science, 7*, 205-267.

Block, N. (1978). Troubles with functionalism. In W. Savage (Ed.), *Minnesota Studies in the Philosophy of Science* (Vol. IX, pp. 261-325). Minneapolis: University of Minnesota Press.

Block, N. (1980). Introduction: What is functionalism? In N. Block (Ed.), *Readings in philosophy of psychology* (Vol. 1, pp. 171-184). Cambridge, MA: Harvard University Press.

Block, N. (1981). Psychologism and behaviorism. *Philosophical Review, 90*, 5-43.

Block, N. (1996). What is functionalism? *The Encyclopaedia of Philosophy Supplement*. Retrieved from http://www.nyu.edu/gsas/dept/philo/faculty/block/papers/functionalism.pdf

Bloomfield, L. (1933). *Language*. New York: Holt, Rinehart and Winston.

Boogerd, F. C., Bruggeman, F. J., Richardson, R. C., Stephan, A., & Westerhoff, H. V. (2005). Emergence and its place in nature: A case study of biochemical networks. *Synthese, 145*, 131-164.

Boyd, R. (1999). Kinds, complexity, and multiple realization. *Philosophical Studies, 95*, 67-98.

Braitenberg, V. (1984). *Vehicles, experiments in synthetic psychology*. Cambridge, MA: MIT Press.

Brentano, F. C. (1874). *Psychologie vom empirischen Standpunkt* (Vol. 1). Leipzig: Duncker & Humblot.

Brooks, R. (1997). Intelligence without representation. In J. Haugeland (Ed.), *Mind design II: Philosophy, psychology, artificial intelligence* (pp. 395-420). Cambridge, MA: MIT Press.

Brooks, R. (1999). *Cambrian intelligence: The early history of the new AI*. Cambridge, MA: MIT Press.

Bruner, J. (1985). *Actual minds, possible worlds*. Cambridge, MA: Harvard University Press.

Burge, T. (1979). Individualism and the mental. *Midwest Studies in Philosophy, 4*, 73-121.

Burge, T. (1982). Other bodies. In A. Woodfield (Ed.), *Thought and object* (pp. 97-120). New York: Oxford University Press.

Burge, T. (1986). Individualism and psychology. *Philosophical Review, 95*, 3-45.

Buss, D. M. (2004). *Evolutionary psychology: The new science of the mind* (2nd ed.). Boston: Allyn and Bacon.

Butler, K. (1998). *Internal affairs: A critique of externalism in the philosophy of mind*. Dordrecht: Kluwer.

Campbell, K. (1984). *Body and mind* (2nd ed.). Notre Dame, IN: University of Notre Dame Press.

Carruthers, M. (1998). *The craft of thought: Meditation, rhetoric, and the making of images, 400-1200*. New York: Cambridge University Press.

Casati, R., & Tappolet, C. (Eds.). (1998). *European Review of Philosophy 3: Response-dependence*. Stanford, CA: CSLI Publications.

Chalmers, D. J. (1996). *The conscious mind: In search of a fundamental theory*. New York: Oxford University Press.

Chalmers, D. (2008). Foreword to Andy Clark's "Supersizing the Mind". In A. Clark (Ed.), *Supersizing the mind: Embodiment, action, and cognitive extension* (pp. vii-xiv). New York: Oxford University Press.

Chambers, D., & Reisberg, D. (1985). Can mental images be ambiguous? *Journal of Experimental Psychology: Human Perception and Performance, 11*, 317-328.

Chemero, A. (2009). *Radical embodied cognitive science*. Cambridge, MA: MIT Press.

Chenn, A. (1997). Development of the cerebral cortex. In W. M. Cowan, S. L. Zipursky & T. M. Jessell (Eds.), *Molecular and cellular approaches to neural development* (pp. 440-473). New York: Oxford University Press.

Chiel, H. J., & Beer, R. D. (1997). The brain has a body: Adaptive behavior emerges from interactions of nervous system, body and environment. *Trends in Neuroscience, 20*, 553-557.

Chiel, H. J., Beer, R. D., & Gallagher, J. C. (1999). Evolution and analysis of model CPGs for walking I. Dynamical modules. *Journal of Computational Neuroscience, 7*, 99-118.

Chisholm, R. M. (1957). *Perceiving: A philosophical study*. Ithaca, NY: Cornell University Press.

Chomsky, N. (1986). *Knowledge of language: Its nature, origins, and use*. New York: Praeger.

Churchland, P. M. (1989). *A neurocomputational perspective: The nature of mind and the structure of science*. Cambridge, MA: MIT Press.

Churchland, P. S., Ramachandran, V. S., & Sejnowski, T. J. (1994). A critique of pure vision. In C. Koch & J. L. Davis (Eds.), *Large-scale neuronal theories of the brain* (pp. 23-60). Cambridge, MA: MIT Press.

Churchland, P. S., & Sejnowski, T. J. (1992). *The computational brain*. Cambridge, MA: MIT Press.

Clancey, W. J. (1997). *Situated cognition: On human knowledge and computer representations*. New York: Cambridge University Press.

Clancey, W. J. (2009). Scientific antecedents of situated cognition. In M. Aydede & P. Robbins (Eds.), *The Cambridge Handbook of Situated Cognition* (pp. 11-34). New York: Cambridge University Press.

Clark, A. (1989). *Microcognition: Philosophy, cognitive science, and parallel distributed processing*. Cambridge, MA: MIT Press.

Clark, A. (1997). *Being there: Putting brain, body, and world together again*. Cambridge, MA: MIT Press.

Clark, A. (1998a). Magic words: How language augments human computation. In P. Carruthers & J. Boucher (Eds.), *Language and thought: Interdisciplinary themes* (pp. 162-183). New York: Cambridge University Press.

Clark, A. (1998b). Twisted tales: Causal complexity and cognitive scientific explanation. *Minds and Machines, 8*, 79-99.
Clark, A. (1998c). Embodied, situated, and distributed cognition. In W. Bechtel, G. Graham & D. A. Balota (Eds.), *A companion to cognitive science* (pp. 506-517). Malden, MA: Blackwell.
Clark, A. (1998d). Time and mind. *Journal of Philosophy, 95*, 354-376.
Clark, A. (2001a). *Mindware: An introduction to the philosophy of cognitive science*. New York: Oxford University Press.
Clark, A. (2001b). Reasons, robots and the extended mind. *Mind and Language, 16*, 121-145.
Clark, A. (2002). Minds, brains, and tools. In H. Clapin (Ed.), *Philosophy of mental representation* (pp. 66-90). New York: Oxford University Press.
Clark, A. (2003). *Natural-born cyborgs: Minds, technologies, and the future of human intelligence*. New York: Oxford University Press.
Clark, A. (2005a). Intrinsic content, active memory and the extended mind. *Analysis, 65*, 1-11.
Clark, A. (2005b). Beyond the flesh: Some lessons from a mole cricket. *Artificial Life, 11*, 233-244.
Clark, A. (2006a). Material symbols. *Philosophical Psychology, 19*, 291-307.
Clark, A. (2006b). Author's reply to symposion on "Natural-Born Cyborgs". *Metascience, 13*, 169-181.
Clark, A. (2007). Curing cognitive hiccups. *Journal of Philosophy, 104*, 163-192.
Clark, A. (2008). *Supersizing the mind: Embodiment, action, and cognitive extension*. New York: Oxford University Press.
Clark, A. (2009). Spreading the joy? Why the machinery of consciousness is (probably) still in the head. *Mind, 118*, 963-993.
Clark, A. (2010). Memento's revenge: The extended mind, extended. In R. Menary (Ed.), *The extended mind* (pp. 43-66). Cambridge, MA: MIT Press.
Clark, A., & Chalmers, D. (1998). The extended mind. *Analysis, 58*, 7-19.
Clark, A., & Grush, R. (1999). Towards a cognitive robotics. *Adaptive Behavior, 7*, 5-16.
Clark, A., & Karmiloff-Smith, A. (1993). The cognizer's innards: A psychological and philosophical perspective on the development of thought. *Mind and Language, 8*, 487-519.
Clark, A. G. (1994). Beliefs and desires incorporated. *Journal of Philosophy, 91*, 404-425.
Colby, K. M. (1975). *Artificial paranoia*. New York: Pergamon.
Coprophagia. Retrieved October 25, 2010, from Wikipedia, the free encylopedia: http://en.wikipedia.org/wiki/Coprophagia

Cosmides, L. (1989). The logic of social exchange: Has natural selection shaped how humans reason? Studies with the Wason selection task. *Cognition, 31*, 187-276.

Cosmides, L., & Tooby, J. (1994). Origins of domain-specificity: The evolution of functional organization. In L. Hirschfeld & S. Gelman (Eds.), *Mapping the mind: Domain-specificity in cognition and culture.* New York: Cambridge University Press.

Cranium Command. Retrieved March 29, 2007, from Wikipedia, the free encylopedia: http://en.wikipedia.org/wiki/Cranium_Command

Craver, C. (2007). *Explaining the brain: Mechanisms and the mosaic unity of neuroscience.* Oxford: Oxford University Press.

Craver, C. F., & Bechtel, W. (2006). Mechanism. In S. Sarkar & J. Pfeifer (Eds.), *Philosophy of science: An encyclopedia* (pp. 469-478). New York: Routledge.

Cruse, H., Dean, J., Dürr, V., Kindermann, T., Schmitz, J., & Schumm, M. (2002). A decentralized, biologically based network for autonomous control of (hexapod) walking. In J. Ayers, J. L. Davis & A. Rudolph (Eds.), *Neurotechnology for biomimetic robots* (pp. 383-400). Cambridge, MA: MIT Press.

Cruse, H., Dean, J., & Ritter, H. (1998). *Die Erfindung der Intelligenz oder Können Ameisen denken?* München: H.C. Beck.

Cruse, H., Kindermann, T., Schumm, M., Dean, J., & Schmitz, J. (1998). Walknet - A biologically inspired network to control six-legged walking. *Neural Networks, 11*, 1435-1447.

Cummins, R. (1983). *The nature of psychological explanation.* Cambridge, MA: MIT Press.

Dartnall, T. (2006). We have always been... cyborgs. *Metascience, 13*, 139-148.

Davidson, D. (1963). Actions, reasons, and causes. *The Journal of Philosophy, 60*, 685-700.

Davidson, D. (1985). Reply to Quine on events. In E. LePore & B. P. McLaughlin (Eds.), *Actions and events: Perspectives on the philosophy of Donald Davidson* (pp. 172-176). Oxford: Blackwell.

Davidson, D. (1987). Knowing one's mind. *Proceedings and Addresses of the American Philosophical Association, 60*, 441-458.

Dawkins, R. (1976). *The selfish gene.* New York: Oxford University Press.

Dawkins, R. (1982). *The extended phenotype.* Oxford: Oxford University Press.

De la Rocha, O. (1985). The reorganization of arithmetic practice in the kitchen. *Anthropology & Education Quarterly, 16*, 171-213.

De Vignemoent, F. (2007). Habeas corpus: The sense of ownership of one's own body. *Mind and Language, 22*, 427-449.

Deacon, T. W. (1997). *The symbolic species: The co-evolution of language and the brain* (1st ed.). New York: W.W. Norton.

Decety, J., & Grèzes, J. (1999). Neural mechanisms subserving the perception of human actions. *Trends in Cognitive Sciences, 3*, 172-178.
Dehaene, S. (1997). *The number sense: How the mind creates mathematics*. New York: Oxford University Press.
Dehaene, S. (2001). Precis of "The number sense". *Mind and Language, 16*, 16-36.
Dehaene, S., & Cohen, L. (1991). Two mental calculation systems: A case study of severe acalculia with preserved approximation. *Neuropsychologia, 29*, 1045-1074.
Dehaene, S., & Cohen, L. (1995). Towards an anatomical and functional model of number processing. *Mathematical Cognition, 1*, 83-120.
Dehaene, S., & Cohen, L. (1997). Cerebral pathways for calculation: Double dissociation between rote verbal and quantitative knowledge of arithmetic. *Cortex, 33*, 219-250.
Dehaene, S., Spelke, E., Pinel, P., Stanescu, R., & Tviskin, S. (1999). Sources of mathematical thinking: Behavioral and brain imaging evidence. *Science, 284*, 970–974.
Delazer, M., & Benke, T. (1997). Arithmetic facts without meaning. *Cortex, 33*, 697-710.
Dennett, D. (1971). Intentional systems. *Journal of Philosophy, 68*, 87-106.
Dennett, D. (1978). *Brainstorms: Philosophical essays on mind and psychology*. Montgomery, VT: Bradford Books.
Dennett, D. (1987). Fast Thinking. In D. Dennett, *The Intentional Stance* (pp. 324-337). Cambridge, MA: MIT Press.
Dennett, D. (1991a). *Consciousness explained*. Boston: Little, Brown and Co.
Dennett, D. (1991b). Real Patterns. *Journal of Philosophy, 88*, 27-51.
Dennett, D. (1993). Labeling and learning. *Mind and Language, 8*, 540-548.
Dennett, D. (1995). *Darwin's dangerous idea: Evolution and the meanings of life*. New York: Simon & Schuster.
Dennett, D. (1996). *Kinds of minds: Toward an understanding of consciousness*. New York: Basic Books.
Dennett, D. (2000). Making tools for thinking. In D. Sperber (Ed.), *Metarepresentations: A multidisciplinary perspective*. Oxford: Oxford University Press.
Descartes, R. (1996). *Meditations on First Philosophy* (J. Cottingham, Trans.). Cambridge, UK: Cambridge University Press.
Devitt, M. (1990). On removing puzzles about belief ascription. *Pacific Philosophical Quarterly, 71*, 165-181.
Devitt, M. (1991). *Realism and truth* (2nd ed.). Oxford: Blackwell.
Di Paolo, E. A. (2005). Autopoiesis, adaptivity, teleology, agency. *Phenomenology and the Cognitive Sciences, 4*, 97-125.

Dickens, W., & Flynn, J. R. (2001). Heritability estimates versus large environmental effects: The IQ paradox resolved. *Psychological Review, 108*, 346-369.

Donald, M. (1991). *Origins of the modern mind: Three stages in the evolution of culture and cognition*. Cambridge, MA: Harvard University Press.

Dretske, F. (1981). *Knowledge and the flow of information*. Cambridge, MA: MIT Press.

Dretske, F. (1986). Misrepresentation. In R. J. Bogdan (Ed.), *Belief: Form, content, and function* (pp. 17-36). Oxford: Clarendon Press.

Dretske, F. (1991). *Explaining behavior: Reasons in a world of causes*. Cambridge, MA: MIT Press.

Dretske, F. (1994). If you can't make one, you don't know how it works. In P. A. French, T. E. Uehling & H. K. Wettstein (Eds.), *Philosophical naturalism* (pp. 468-482). Notre Dame, IN: University of Notre Dame Press.

Dror, I., & Harnad, S. (2008). *Cognition distributed: How cognitive technology extends our minds*. Amsterdam: John Benjamins.

Dürr, V., Krause, A., Schmitz, J., & Cruse, H. (2003). Neuroethological concepts and their transfer to walking machines. *International Journal of Robotics Research, 22*, 151-167.

Dürr, V., Schmitz, J., & Cruse, H. (2004). Behavior-based modelling of hexapod locomotion: linking biology and technical application. *Arthropod Structure & Development, 33*, 237-250.

Egan, F. (1992). Individualism, computation and perceptual content. *Mind, 101*, 443-459.

Ehret, G. (1987). Categorical perception of sound signals: Facts and hypotheses from animal studies. In S. Harnad (Ed.), *Categorical perception: The groundwork of cognition* (pp. 301-331). Cambridge, UK: Cambridge University Press.

Eliasmith, C. (1996). The third contender: A critical examination of the dynamicist theory of cognition. *Philosophical Psychology, 9*, 441-463.

Elman, J. L., Bates, E., Johnson, M. H., Karmiloff-Smith, A., Parisi, D., & Plunkett, K. (1996). *Rethinking innateness: A connectionist perspective on development*. Cambridge, MA: MIT Press.

Erlhagen, W., & Schöner, G. (2002). Dynamic field theory of motor preparation. *Psychological Review, 109*, 545-572.

Evans, J. S. B. T., Newstead, S. E., & Byrne, R. M. J. (1993). *Human reasoning: The psychology of deduction*. Hove, UK: Lawrence Erlbaum Associates.

Ferreiro, E. (1991). Literacy development: A psychogenetic perspective. In D. R. Olson & N. Torrance (Eds.), *Literacy and orality* (pp. 217-228). Cambridge, MA: Harvard University Press.

Ferreiro, E. (1994). Written representations of language. In D. Keller-Cohen (Ed.), *Literacy: Interdisciplinary conversations* (pp. 115-128). Cresskill, NJ: Hampton Press.

Fine, K. (1975). Vagueness, truth and logic. *Synthese, 54*, 235-259.

Flavell, J. H., & Wellman, H. M. (1977). Metamemory. In R. V. Kail & J. W. Hagen (Eds.), *Perspectives on the development of memory and cognition* (pp. 3-33). Hillsdale, NJ: Lawrence Erlbaum Associates.

Flor, N. V., & Hutchins, E. (1992). Analyzing distributed cognition in software teams: A case study of collaborative programming during adaptive software maintenance. In J. Koenemann-Belliveau, T. Moher & S. Robertson (Eds.), *Empirical Studies of Programmers: Fourth Workshop* (pp. 36-64). Norwood, NJ: Ablex.

Flynn, J. R. (1984). The mean IQ of Americans: Massive gains 1932 to 1978. *Psychological Bulletin, 95*, 29-51.

Flynn, J. R. (1994). IQ gains over time. In R. J. Sternberg (Ed.), *Encyclopedia of human intelligence* (pp. 617-623). New York: Macmillan.

Fodor, J. (1974). Special sciences (or: the disunity of science as a working hypothesis). *Synthese: An International Journal for Epistemology, Methodology and Philosophy of Science, 28*, 97-115.

Fodor, J. (1975). *The language of thought*. Cambridge, MA: Harvard University Press.

Fodor, J. (1980). Methodological solipsism considered as a research strategy in cognitive science. *Behavioral and Brain Sciences, 3*, 63-109.

Fodor, J. (1983). *Modularity of mind: An essay on faculty psychology*. Cambridge, MA: MIT Press.

Fodor, J. (1985). Fodor's guide to mental representation: The intelligent auntie's vade-mecum. *Mind, 94*, 76-100.

Fodor, J. (1987). *Psychosemantics: The problem of meaning in the philosophy of mind*. Cambridge, MA: MIT Press.

Fodor, J. (1990). *A theory of content and other essays*. Cambridge, MA: MIT Press.

Fodor, J. (1991). A modal argument for narrow content. *Journal of Philosophy, 88*, 1-22.

Fodor, J. (1994). *The elm and the expert: Mentalese and its semantics*. Cambridge, MA: MIT Press.

Fodor, J. (1997). Special sciences: Still autonomous after all these years. *Philosophical Perspectives, 11*, 149-163.

Fowler, C. A., & Turvey, M. T. (1978). Skill acquisition: An event approach with special reference to searching for the optimum of a function of several variables. In G. E. Stelmach (Ed.), *Information processing in motor control and learning* (pp. 2-38). New York: Academic Press.

Frankish, K. (1998a). Natural language and virtual belief. In P. Carruthers & J. Boucher (Eds.), *Language and thought: Interdisciplinary themes* (pp. 248-269). Cambridge, UK: Cambridge University Press.

Frankish, K. (1998b). A matter of opinion. *Philosophical Psychology, 11*, 423-443.

Franks, B. (1995). On explanation in cognitive science: Competence, idealization, and the failure of the classical cascade. *British Journal for the Philosophy of Science, 46*, 475-502.

Freud, S. (1955). Eine Schwierigkeit der Psychoanalyse. In S. Freud, A. Freud & L. Veszy-Wagner (Eds.), *Gesammelte Werke* (Vol. 12, pp. 3-12). London: Imago.

Gallagher, S. (2000). Philosophical conceptions of the self: Implications for cognitive science. *Trends in Cognitive Sciences, 4*, 14-21.

Gallagher, S. (2009). Philosophical antecedents to situated cognition. In M. Aydede & P. Robbins (Eds.), *The Cambridge Handbook of Situated Cognition* (pp. 35-51). New York: Cambridge University Press.

Gallagher, S., & Crisafi, A. (2009). Mental institutions. *Topoi, 28*, 45-51.

Garbis, C., & Waern, Y. (1999). Team co-ordination and communication in a rescue command staff: The role of public representations. *Le Travail Humain, 62*, 273-291.

Geertz, C. (1983). *Local knowledge: Further essays in interpretive anthropology*. New York: Basic Books.

Gell-Mann, M. (1994). *The quark and the jaguar: Adventures in the simple and the complex*. New York: W. H. Freeman.

Gertler, B. (2007). Overextending the mind? In B. Gertler & L. Shapiro (Eds.), *Arguing about the mind*. New York: Routledge.

Gigerenzer, G. (2000). *Adaptive thinking: Rationality in the real world*. New York: Oxford University Press.

Gigerenzer, G., Todd, P. M., & Group, A. R. (1999). *Simple heuristics that make us smart*. New York: Oxford University Press.

Gillet, C. (2002). The dimensions of realization: A critique of the standard view. *Analysis, 62*, 316-323.

Gillet, C. (2003). The metaphysics of realization, multiple realizability and the special sciences. *Journal of Philosophy, 100*, 591-603.

Glenberg, A. M. (1997). What memory is for. *Behavioral and Brain Sciences, 20*, 1-55.

Godzich, W., & Kittay, J. (1987). *The emergence of prose: An essay in prosaics*. Minneapolis: University of Minnesota Press.

Goldstone, R. L., Landy, D., & Son, J. Y. (2008). A well grounded education: The role of perception in science and mathematics. In M. de Vega, A. Glenberg & A. Graesser (Eds.), *Symbols and embodiment: Debates on meaning and cognition* (pp. 327-355). Oxford: Oxford University Press.

Goodale, M. A., & Milner, A. D. (2004). *Sight unseen: An exploration of conscious and unconscious vision*. Oxford: Oxford University Press.
Goody, J. (1977). *The domestication of the savage mind*. Cambridge, UK: Cambridge University Press.
Goody, J. (2000). *The power of the written tradition*. Washington: Smithsonian Institution Press.
Goody, J., & Watt, I. (1963). The consequences of literacy. *Comparative Studies in Society and History, 5*, 304-345.
Grafton, S. T., Fadiga, L., Arbib, M. A., & Rizzolatti, G. (1997). Premotor cortex activation during observation and naming of familiar tools. *Neuroimage, 6*, 231-236.
Graves, D. (1975). An examination of the writing process of seven year old children. *Research in the Teaching of English, 9*, 227-241.
Greeno, J. G., & Moore, J. L. (1993). Situativity and symbols: Response to Vera and Simon. *Cognitive Science, 17*, 49-59.
Griffiths, P. E., & Stotz, K. (2000). How the mind grows: A developmental perspective on the biology of cognition. *Synthese, 122*, 29-51.
Grillner, S., McClellan, A., & Perret, C. (1981). Entrainment of the spinal pattern generators for swimming by mechano-sensitive elements in the lamprey spinal cord in vitro. *Brain Research, 217*, 380-386.
Grush, R. (2003). In defence of some "Cartesian" assumptions concerning the brain and its operation. *Biology and Philosophy, 18*, 53-93.
Grush, R. (2004). The emulation theory of representation: Motor control, imagery, and perception. *Behavioral and Brain Sciences, 27*(3), 377-442.
Haken, H. (1996). *Principles of brain functioning: A synergetic approach to brain activity, behavior and cognition*. Berlin: Springer.
Haken, H., Kelso, J. A. S., & Bunz, H. (1985). A theoretical model of phase transitions in human hand movements. *Biological Cybernetics, 51*, 347-356.
Harman, G. (1987). (Non-solipsistic) Conceptual role semantics. In E. LePore (Ed.), *New directions in semantics* (pp. 55-81). London: Academic Press.
Harnad, S. (Ed.). (1987). *Categorical perception: The groundwork of cognition*. Cambridge, UK: Cambridge University Press.
Harnad, S. (1990). The symbol grounding problem. *Physica D, 42*, 335-346.
Harnad, S. (1991). Other bodies, other minds: A machine incarnation of an old philosophical problem. *Minds and Machines, 1*, 43-54.
Harnad, S. (2005). Distributed processes, distributed cognizers and collaborative cognition. *Pragmatics and Cognition, 13*, 501-514.
Harris, R. (1986). *The origin of writing*. London: Duckworth.
Hart, J. T. (1967). Memory and the memory-monitoring process. *Journal of Verbal Learning and Verbal Behavior, 6*, 685-691.

Haugeland, J. (1998). Mind embodied and embedded. In J. Haugeland (Ed.), *Having thought: Essays in the metaphysics of mind* (pp. 207-240). Cambridge, MA: Harvard University Press.

Havelock, E. A. (1963). *Preface to Plato*. Cambridge, MA: Belknap Press.

Havelock, E. A. (1976). *Origins of western literacy*. Toronto: Ontario Institute for Studies in Education.

Hearst, E., & Knott, J. (2008). *Blindfold chess: History, psychology, techniques, champions, world records, and important games*. Jefferson, NC: Macfarland.

Heidegger, M. (1927/1977). *Sein und Zeit*. Frankfurt/Main: Vittorio Klostermann.

Herken, R. (1995). *The Universal Turing Machine. A half-century survey*. Berlin: Springer.

Hermer-Vazquez, L., Spelke, E., & Katsnelson, A. (1999). Sources of flexibility in human cognition: Dual-task studies of space and language. *Cognitive Psychology, 39*, 3-36.

Hollan, J., Hutchins, E., & Kirsh, D. (2000). Distributed cognition: Toward a new foundation for human-computer interaction research. *ACM Transactions on Computer-Human Interaction, Special issue on human-computer interaction in the new millennium, Part 2, 7*, 174-196.

Holyoak, K. J., & Thagard, P. (1995). *Mental leaps: Analogy in creative thought*. Cambridge, MA: MIT Press.

Homer, B., & Olson, D. R. (1999). Literacy and children's conception of words. *Written Language and Literacy, 2*, 113-137.

Horgan, T. (1993). Nonreductive materialism and the explanatory autonomy of psychology. In S. Wagner & R. Warner (Eds.), *Naturalism: A critical appraisal* (pp. 295-320). Notre Dame, IN: University of Notre Dame Press.

Horgan, T., & Tienson, J. (2002). The intentionality of phenomenology and the phenomenology of intentionality. In D. J. Chalmers (Ed.), *Philosophy of Mind*. Oxford: Oxford University Press.

Horowitz, R., & Olson, D. R. (2007). Texts that talk: The special and peculiar nature of classroom discourse and the crediting of sources. In R. Horowitz (Ed.), *Talking texts: How speech and writing interact in school learning* (pp. 55-90). New York: Lawrence Erlbaum Associates.

Horst, S. (2011). The computational theory of mind. In E. N. Zalta (Ed.), *The Stanford Encyclopedia of Philosophy* (Spring 2011 ed.): URL = <http://plato.stanford.edu/archives/spr2011/entries/computational-mind/>.

Hume, D. (1975). *Enquiries concerning human understanding and concerning the principles of morals* (L. A. Selby-Bigge and P. H. Nidditch ed.). Oxford: Clarendon Press.

Hurley, S. L. (1998a). Vehicles, contents, conceptual structure, and externalism. *Analysis, 58*, 1-6.

Hurley, S. L. (1998b). *Consciousness in action*. Cambridge, MA: Harvard University Press.
Husbands, P., Harvey, I., & Cliff, D. (1995). Circle in the round: State space attractor for evolved sight robots. *Journal of Robotics and Autonomous Systems, 15*, 83-106.
Hutchins, E. (1995a). *Cognition in the wild*. Cambridge, MA: MIT Press.
Hutchins, E. (1995b). How a cockpit remembers its speed. *Cognitive Science, 19*, 265-288.
Hutchins, E., & Klausen, T. (1996). Distributed cognition in an airline cockpit. In Y. Engestrom & D. Middleton (Eds.), *Cognition and communication at work* (pp. 15-34). New York: Cambridge University Press.
Huygens, C. (1673). Horologium oscillatorium. Paris: F. Muguet.
Iida, F., & Pfeifer, R. (2004). Self-stabilization and behavioral diversity of embodied adaptive locomotion. In F. Iida, R. Pfeifer, L. Steels & Y. Kuniyoshi (Eds.), *Embodied artificial intelligence* (pp. 119-129). Berlin: Springer.
Jablonka, E., & Lamb, M. (2005). *Evolution in four dimensions: Genetic, epigenetic, behavioral, and symbolic variation in the history of life*. Cambridge, MA: MIT Press.
Jackendoff, R. (1996). How language helps us think. *Pragmatics and Cognition, 4*, 1-24.
Jackson, F. (1998). *From metaphysics to ethics: A defence of conceptual analysis*. Oxford: Oxford University Press.
Johnson-Laird, P. N. (1983). *Mental models: Towards a cognitive science of language, inference, and consciousness*. Cambridge, MA: Harvard University Press.
Jonassen, D. H., Mayes, T., & McAleese, R. (1991). A manifesto for a constructivist approach to uses of technology in higher education. In T. M. Duffy, J. Lowyk & D. H. Jonassen (Eds.), *Designing constructivist learning environments* (pp. 232-247). Heidelberg: Springer.
Kallestrup, J. (2010). *Semantic Externalism*. New York: Routledge.
Karmiloff-Smith, A. (1992). *Beyond modularity: A developmental perspective on cognitive science*. Cambridge, MA: MIT Press.
Karmiloff-Smith, A. (1994). Precis of "Beyond modularity: A developmental perspective on cognitive science". *Behavioral and Brain Sciences, 17*, 693-745.
Karmilloff-Smith, A., Grant, J., Sims, K., Jones, M.-C., & Cuckle, P. (1996). Rethinking metalinguistic awareness: Representing and accessing knowledge about what counts as a word. *Cognition, 58*, 197-219.
Keijzer, F. (2001). *Representation and behavior*. Cambridge, MA: MIT Press.

Kelso, J. A. S. (1986). Pattern formation in speech and limb movements involving many degrees of freedom. *Experimental Brain Research Series, 15*, 105-128.

Kelso, J. A. S. (1995). *Dynamic patterns: The self-organization of brain and behavior*. Cambridge, MA: MIT Press.

Kim, J. (1993). *Supervenience and mind.* New York: Cambridge University Press.

Kim, J. (1996). *Philosophy of mind.* Boulder, CO: Westview Press.

Kim, J. (2001). Lonely souls: Causality and substance dualism. In K. Corcoran (Ed.), *Soul, body, and survival: Essays on the metaphysics of human persons* (pp. 30-43). Ithaca, NY: Cornell University Press.

Kirsh, D. (1995). The intelligent use of space. *Artificial Intelligence, 72*, 1-52.

Kirsh, D. (2009). Problem solving and situated cognition. In M. Aydede & P. Robbins (Eds.), *The Cambridge Handbook of Situated Cognition* (pp. 264-306). New York: Cambridge University Press.

Kirsh, D., & Maglio, P. (1994). On distinguishing epistemic from pragmatic action. *Cognitive Science, 18*, 513-549.

Kiverstein, J., & Clark, A. (2009). Introduction: Mind embodied, embedded, enacted: One church or many? *Topoi, 28*, 1-7.

Knobe, J., & Prinz, J. (2008). Intuitions about Consciousness: Experimental Studies. *Phenomenology and the Cognitive Sciences, 7*, 67-83.

Köhler, W. (1925). *The mentality of apes*. London: Routledge & Kegan Paul.

Kornblith, H. (1993). *Inductive inference and its natural ground: An essay in naturalistic epistemology*. Cambridge, MA: MIT Press.

Kristan, W. B., Skalak, R., Wilson, R. J. A., Skierczynski, B. A., Murray, J. A., Eisenhart, F. J. (1996). Biomechanics of hydroskeletons: Lessons learned from studies of crawling in the medicinal leech. In J. M. Winters & P. E. Crago (Eds.), *Biomechanics and Neural Control of Movement - 9th Engineering Foundation Conference* (pp. 206-218).

Kuhn, T. S. (1970). *The structure of scientific revolutions* (2nd ed.). Chicago: University of Chicago Press.

Lakoff, G., & Núñez, R. E. (2000). *Where mathematics comes from: How the embodied mind brings mathematics into being*. New York: Basic Books.

Landy, D., & Goldstone, R. (2007). How abstract is symbolic thought? *Journal of Experimental Psychology: Learning, Memory, and Cognition, 33*, 720-733.

Langacker, R. W. (1987). *Foundations of cognitive grammar: Theoretical Prerequisites* (Vol. 1). Stanford, CA: Stanford University Press.

Langacker, R. W. (1991). *Foundations of cognitive grammar: Descriptive application* (Vol. 2). Stanford, CA: Stanford University Press.

Larkin, J. H., & Simon, H. A. (1987). Why a diagram is (sometimes) worth ten thousand words. *Cognitive Science, 11*, 65-99.

Latour, B. (1986). Visualization and cognition: Thinking with eyes and hands. *Knowledge and Society, 6*, 1-40.

Lau, J., & Deutsch, M. (2010). Externalism about mental content. In E. N. Zalta (Ed.), *The Stanford Encyclopedia of Philosophy* (Fall 2010 ed.): URL = <http://plato.stanford.edu/archives/fall2010/entries/content-externalism/>.

Lettvin, J. Y., Maturana, H. R., McCulloch, W. S., & Pitts, W. H. (1959). What the frog's eye tells the frog's brain. *Proceedings of the IRE, 47*, 1940-1959.

Levin, J. (2010). Functionalism. In E. N. Zalta (Ed.), *The Stanford Encyclopedia of Philosophy* (Summer 2010 ed.): URL = <http://plato.stanford.edu/archives/sum2010/entries/functionalism/>.

Levine, J. (1983). Materialism and qualia: The explanatory gap. *Pacific Philosophical Quarterly, 64*, 354-361.

Lewis, D. (1970). How to define theoretical terms. *The Journal of Philosophy, 67*, 427-446.

Lewis, D. (1972). Psychophysical and theoretical identifications. *Australasian Journal of Philosophy, 50*, 249-258.

Lewis, D. (1973). Causation. *Journal of Philosophy, 70*, 556-567.

Lewis, D. (1983). New work for a theory of universals. *Australasian Journal of Philosophy, 61*, 343-377.

Lewis, D. (1986). Postscripts to "Causation". In D. Lewis (Ed.), *Philosophical papers* (Vol. 2, pp. 172-213). New York: Oxford University Press.

Linden, D. (2007). *The accidental mind: How brain evolution has given us love, memory, dreams, and God*. Cambridge, MA: Belknap Press.

Linell, P. (2005). *The written language bias in linguistics: Its nature, origins and transformations*. London: Routledge.

Locke, J. (1975). *An essay concerning human understanding* (P. H. Nidditch ed.). Oxford: Clarendon Press.

Lungarella, M., & Sporns, O. (2005). Information self-structuring: Key principle for learning and development. In C. Yu, H. Zhang & L. B. Smith (Eds.), *Proceedings 2005 IEEE International Conference on Development and Learning* (pp. 25-30).

Luria, A. R. (1976). *Cognitive development: Its cultural and social foundations*. Cambridge, MA: Harvard University Press.

Lycan, W. (1981). Toward a homuncular theory of believing. *Cognition and Brain Theory, 4*, 139-159.

Lycan, W. (1987). *Consciousness*. Cambridge, MA: MIT Press.

MacIver, M. (2009). Neuroethology: From morphological computation to planning. In P. Robbins & M. Aydede (Eds.), *The Cambridge Handbook of Situated Cognition* (pp. 480-504). New York: Cambridge University Press.

Malafouris, L. (2008). Tectonoetic awareness and the extended self. *Philosophical Transactions of the Royal Society B: Biological Sciences, 363*, 1993-2002.

Marcus, G. (2008). *Kluge: The haphazard construction of the human mind*. Boston: Houghton Mifflin Harcourt.

Mareschal, D., Johnson, M. H., Sirois, S., Spratling, M., Thomas, M., & Westermann, G. (2007a). *Neuroconstructivism: How the brain constructs cognition* (Vol. 1). Oxford: Oxford University Press.

Mareschal, D., Sirois, S., & Westermann, G. (2007b). *Neuroconstructivism: Perspectives and prospects* (Vol. 2). Oxford: Oxford University Press.

Marr, D. (1982). *Vision: A computational investigation into the human representation and processing of visual information*. San Francisco: W. H. Freeman.

Marsh, L., & Onof, C. (2008). Introduction to the special issue "Perspectives on social cognition". *Cognitive Systems Research, 9*, 1-4.

Matthen, M. (2005). *Seeing, doing and knowing: A philosophical theory of sense perception*. Oxford: Clarendon.

Maxwell, J. C. (1868). On governors. *Proceedings of the Royal Society, 100*, 270-283.

McClamrock, R. A. (1995). *Existential cognition: Computational minds in the world*. Chicago: University of Chicago Press.

McGinn, C. (1989). *Mental content*. New York: Blackwell.

McKitrick, J. (2003). A case for extrinsic dispositions. *Australasian Journal of Philosophy, 81*, 155-174.

McLaughlin, B. P. (1991). *Dretske and his critics*. Cambridge, MA: Blackwell.

McLuhan, M. (1962). *The Gutenberg galaxy: The making of typographic man*. Toronto: University of Toronto Press.

Melnyk, A. (2003). *A physicalist manifesto: Thoroughly modern materialism*. Cambridge, UK: Cambridge University Press.

Menary, R. (2006). Attacking the bounds of cognition. *Philosophical Psychology, 19*, 329-344.

Menary, R. (2007). *Cognitive integration: Mind and cognition unbounded*. Basingstoke, UK: Palgrave Macmillan.

Millikan, R. G. (1984). *Language, thought, and other biological categories: New foundations for realism*. Cambridge, MA: MIT Press.

Millikan, R. G. (1989). Biosemantics. *Journal of Philosophy, 86*, 281-297.

Millikan, R. G. (1993). *White queen psychology and other essays for Alice*. Cambridge, MA: MIT Press.

Milner, A. D., & Goodale, M. A. (1995). *The visual brain in action*. Oxford: Oxford University Press.

Mindell, D. (2008). *Digital Apollo: Human and machine in the first six lunar landings*. Cambridge, MA: MIT Press.

Mithen, S. J. (1996). *The prehistory of the mind: A search for the origins of art, religion, and science.* London: Thames and Hudson.
Mithen, S. J. (2000). Mind, brain and material culture: An archaeological perspective. In P. Carruthers & A. Chamberlain (Eds.), *Evolution and the human mind: Modularity, language, and meta-cognition* (pp. 207-217). Cambridge, UK: Cambridge University Press.
Monaghan, P., & Stenning, K. (1998). Effects of representational modality and thinking style on learning to solve reasoning problems. In A. Gernsbacher & S. J. Derry (Eds.), *20th Annual Conference of the Cognitive Science Society* (pp. 716-721). Mahwah, NJ: Lawrence Erlbaum Associates.
Monaghan, P., Stenning, K., Oberlander, J., & Sönstrod, C. (1999). Integrating psychometric and computational approaches to individual differences in multimodal reasoning. In M. Hahn & S. C. Stoness (Eds.), *21st Annual Cognitive Science Society Conference* (pp. 405-410). Mahwah, NJ: Lawrence Erlbaum Associates.
Nagel, E. (1961). *The structure of science: Problems in the logic of scientific explanation.* New York: Harcourt, Brace & World.
Nagel, T. (1971). Brain bisection and the unity of consciousness. *Synthese, 22*, 396-413.
Nagel, T. (1974). What is it like to be a bat? *Philosophical Review, 83*, 435-450.
Neander, K. (2009). Teleological theories of mental content. In E. N. Zalta (Ed.), *Stanford Encyclopedia of Philosophy* (Winter 2009 ed.): URL = <http://plato.stanford.edu/archives/win2009/entries/content-teleological/>.
Nelson, T., & Narens, L. (1990). Metamemory: A theoretical framework and new findings. In G. H. Bower (Ed.), *The psychology of learning and motivation: Advances in research and theory* (pp. 125-173). New York: Academic Press.
Newell, A., & Simon, H. (1976). Computer science as empirical inquiry: Symbols and search. *Communications of the ACM, 19*, 113-126.
Nielsen, C. (2001). *Animal evolution: Interrelationships of the living phyla* (2nd ed.). Oxford: Oxford University Press.
Noë, A. (2004). *Action in perception.* Cambridge, MA: MIT Press.
Noë, A. (2009). *Out of our heads: Why you are not your brain and other lessons from the biology of consciousness.* New York: Hill and Wang.
Nolan, C. (2001). Memento [videorecording]. Culver City, NA: Columbia TriStar Home Entertainment.
Norman, D. (1988). *The psychology of everyday things.* New York: Currency-Doubleday.
Norman, D. (1991). Cognitive artifacts. In J. M. Carroll (Ed.), *Designing interaction: Psychology at the human-computer interface* (pp. 17-38). New York: Cambridge University Press.

Norman, D. (1993). *Things that make us smart: Defending human attributes in the age of the machine*. Cambridge, MA: Perseus.

Norman, D. (1999). *The invisible computer*. Cambridge, MA: MIT Press.

Northcut, R. G. (2002). Understanding vertebrate brain evolution. *Integrative and Comparative Biology, 42*, 743-756.

O'Brien, G. (1998). Review symposion: "Being there". *Metascience, 7*, 78-83.

O'Connor, T. (2000). *Persons and causes: The metaphysics of free will*. New York: Oxford University Press.

Odling-Smee, F. J., Laland, K. N., & Feldman, M. W. (2003). *Niche construction: The neglected process in evolution*. Princeton: Princeton University Press.

Olson, D. (1991). Literacy as metalinguistic activity. In D. R. Olson & N. Torrance (Eds.), *Literacy and orality* (pp. 251-270). New York: Cambridge University Press.

Olson, D. (1994). *The world on paper: The conceptual and cognitive implications of writing and reading*. New York: Cambridge University Press.

Olson, D. (1996). Language and literacy: What writing does to language and mind. *Annual Review of Applied Linguistics, 16*, 3-13.

Olson, D. (2001). What writing is. *Pragmatics and Cognition, 9*, 239-258.

Olson, D. (2002). What writing does to the mind. In E. Amsel & J. P. Byrnes (Eds.), *Language, literacy, and cognitive development: The development and consequences of symbolic communication* (pp. 153-165). Mahwah, NJ: Lawrence Erlbaum Associates.

Olson, D. (2007). Self-ascription of intention: Responsibility, obligation, and self-control. *Synthese, 159*, 297-314.

Olson, D. (2008). *Footnotes to Goody: On Goody and his critics*. Paper presented at the ENSSIB in Lyon, France.

Olson, D., & Astington, J. W. (1990). Talking about text: How literacy contributes to thought. *Journal of Pragmatics, 14*, 557-573.

Olson, D., & Torrance, N. (1987). Language, literacy and mental states. *Discourse Processes, 10*, 157-167.

Ong, W. J. (1982). *Orality and literacy: The technologizing of the word*. London: Methuen

Ong, W. J. (1986). Writing is a technology that restructures thought. In G. Baumann (Ed.), *The written word: Literacy in transition* (pp. 23-50). Oxford: Clarendon Press.

Oppy, G., & Dowe, D. (2011). The Turing Test. In E. N. Zalta (Ed.), *The Stanford Encyclopedia of Philosophy* (Spring 2011 ed.): URL = <http://plato.stanford.edu/archives/spr2011/entries/turing-test/>.

O'Regan, K. (1992). Solving the "real" mysteries of visual perception: The world as an outside memory. *Canadian Journal of Psychology, 46*, 461-488.

O'Regan, K., & Noë, A. (2001). A sensorimotor account of vision and visual consciousness. *Behavioral and Brain Sciences, 24*, 939-1011.

Oyama, S. (1985). *The ontogeny of information: Developmental systems and evolution*. Cambridge, UK: Cambridge University Press.

Oyama, S., Griffiths, P. E., & Gray, R. D. (2001). *Cycles of contingency: Developmental systems and evolution*. Cambridge, MA: MIT Press.

Pask, G. (1975). *Conversation, cognition and learning: A cybernetic theory and methodology*. Amsterdam: Elsevier.

Pauen, M. (2007). *Was ist der Mensch? Die Entdeckung der Natur des Geistes*. München: Deutsche Verlags-Anstalt.

Paul, C. (2004). Morphology and Computation. In S. Schaal, A. J. Ijspeert, A. Billard & S. Vijayakumar (Eds.), *Proceedings of the International Conference on the Simulation of Adaptive Behaviour* (pp. 33-38). Los Angeles: MIT Press.

Paulin, M. G. (2005). Evolutionary origins and principles of distributed neural computation for state estimation and movement control in vertebrates. *Complexity, 10*, 56-65.

Perry, M. (2003). Distributed cognition. In J. M. Carroll (Ed.), *HCI models, theories, and frameworks: Toward an interdisciplinary science* (pp. 193-223). San Francisco: Morgan Kaufmann.

Pettit, P. (1991). Realism and response-dependence. *Mind, 100*, 587-626.

Pfeifer, R., Bongard, J., & Grand, S. (2007). *How the body shapes the way we think: A new view of intelligence*. Cambridge, MA: MIT Press.

Pfeifer, R., Iida, F., & Bongard, J. (2005). New robotics: Design principles for intelligent systems. *Artificial Life, Special Issue on New Robotics, Evolution and Embodied Cognition, 11*, 99-120.

Pfeifer, R., Iida, F., & Gómez, G. (2006). Designing intelligent robots – on the implications of embodiment. *Journal of Robotics Society of Japan, 24*, 9-16.

Pfeifer, R., Iida, F., & Gómez, G. (2006). Morphological computation for adaptive behavior and cognition. *International Congress Series, 1291*, 22-29.

Pfeifer, R., & Scheier, C. (1999). *Understanding intelligence*. Cambridge, MA: MIT Press.

Piaget, J. (1955). *The child's construction of reality*. London: Routledge & Paul.

Pietroski, P. M. (2000). *Causing actions*. Oxford: Oxford University Press.

Pinker, S. (1997). *How the mind works*. New York: Norton.

Plato. (1997). *Cratylos* (C. D. C. Reeve, Trans.). Indianapolis: Hackett.

Port, R. F., & Van Gelder, T. (1995). *Mind as motion: Explorations in the dynamics of cognition*. Cambridge, MA: MIT Press.

Premack, D., & Woodruff, G. (1978). Does the chimpanzee have a theory of mind? *Behavioral & Brain Sciences, 1*, 515-526.

Prinz, J., & Clark, A. (2004). Putting concepts to work: Some thoughts for the 21st century. *Mind And Language, 19*, 57-69.

Prior, E., Pargetter, R., & Jackson, F. (1982). Three theses about dispositions. *American Philosophical Quarterly, 3*, 251-257.

Putnam, H. (1960). Minds and Machines. In S. Hook (Ed.), *Dimensions of mind* (pp. 148-180). New York: New York University Press.

Putnam, H. (1975a). The nature of mental states. In H. Putnam, *Mind, language, and reality* (pp. 429-440). Cambridge, UK: Cambridge University Press.

Putnam, H. (1975b). The meaning of "meaning". In H. Putnam, *Mind, language, and reality* (pp. 215-271). Cambridge, UK: Cambridge University Press.

Putnam, H. (1981). *Reason, truth, and history*. Cambridge, UK: Cambridge University Press.

Putnam, H. (1988). *Representation and reality*. Cambridge, MA: MIT Press.

Quartz, S. R. (1999). The constructivist brain. *Trends in Cognitive Sciences, 3*, 48-57.

Quartz, S. R., & Sejnowksi, T. J. (1997). The neural basis of cognitive development: A constructivist manifesto. *Behavioral and Brain Sciences, 20*, 537-596.

Quine, W. v. O. (1953). On what there is. In W. v. O. Quine, *From a logical point of view* (pp. 1-19). Cambridge, MA: Harvard University Press.

Quine, W. v. O. (1960). *Word and object*. Cambridge, MA: MIT Press.

Rawls, J. (2001). *Justice as fairness: A restatement*. Cambridge, MA: Belknap Press.

Rey, G. (1997). *Contemporary philosophy of mind*. Cambridge, MA: Blackwell.

Rey, G. (2002). Searle's misunderstandings of functionalism and Strong AI. In J. Preston & M. Bishop (Eds.), *Views into the Chinese Room: New essays on Searle and Artificial Intelligence* (pp. 201-225). New York: Oxford University Press.

Richerson, P. J., & Boyd, R. (2005). *Not by genes alone: How culture transformed human evolution*. Chicago: University of Chicago Press.

Rips, L. J. (1994). *The psychology of proof: Deductive reasoning in human thinking*. Cambridge, MA: MIT Press.

Robbins, P. (2008). Consciousness and the social mind. *Cognitive Systems Research, 9*, 15-23.

Robbins, P., & Aydede, M. (Eds.). (2009). *The Cambridge Handbook of Situated Cognition*. New York: Cambridge University Press.

Rockwell, W. T. (2005). *Neither brain nor ghost: A nondualist alternative to the mind-brain identity theory*. Cambridge, MA: MIT Press.

Roe, A. W., Pallas, S. L., Hahm, J. O., & Sur, M. (1990). A map of visual space induced into primary auditory cortex. *Science, 250*, 818-820.

Rogers, Y., & Ellis, J. (1994). Distributed cognition: An alternative framework for analysing and explaining collaborative working. *Journal of Information Technology, 9*, 119-128.

Roscher, W. H. (1894). Kentauren. In W. H. Roscher (Ed.), *Ausführliches Lexikon der griechischen und römischen Mythologie* (Vol. 2, 1, pp. 1032-1088). Leipzig: Teubner.

Rosenblatt, F. (1962). *Principles of neurodynamics. Perceptrons and the theory of brain mechanisms*. Washington, DC: Spartan Books.

Rowlands, M. (1999). *The body in mind: Understanding cognitive processes*. Cambridge, UK: Cambridge University Press.

Rowlands, M. (2003). *Externalism: Putting mind and world back together again*. Montreal: McGill-Queen's University Press.

Rowlands, M. (2006). *Body language: Representation in action*. Cambridge, MA: MIT Press.

Rowlands, M. (2008). *Perception: From extended mind to embodied phenomenology*. Paper presented at the Symposion on Embodied Perception held at the Spring 2008 meeting of the Pacific Division of the American Philosophical Association, Pasadena, CA.

Rumelhart, D. E., & McClelland, J. L. (1986). *Parallel distributed processing: Explorations in the microstructure of cognition*. Cambridge, MA: MIT Press.

Rupert, R. (2004). Challenges to the hypothesis of extended cognition. *The Journal of Philosophy, 101*, 389-428.

Rupert, R. (2005). Minding one's cognitive systems: When does a group of minds constitute a single cognitive unit? *Episteme, 1*, 177-188.

Rupert, R. (2006). *Extended cognition as a framework for empirical psychology: The costs outweigh the benefits*. Paper presented at the Central Division Meeting of the American Philosophical Association, Chicago, IL.

Rupert, R. (2009). *Cognitive systems and the extended mind*. New York: Oxford University Press.

Rupert, R. (2010a). Extended cognition and the priority of cognitive systems. *Cognitive Systems Research, 11*, 343-356.

Rupert, R. (2010b). Representation in extended cognitive systems: Does the scaffolding of language extend the mind? In R. Menary (Ed.), *The extended mind* (pp. 325-354). Cambridge, MA: MIT Press.

Russell, S. J., & Norvig, P. (2003). *Artificial Intelligence: A modern approach* (2nd ed.). Upper Saddle River, NJ: Prentice Hall.

Salzman, C. D., & Newsome, W. T. (1994). Neural mechanisms for forming a perceptual decision. *Science, 264*, 231-237.

Sandelands, L. E., & Stablein, R. E. (1987). The concept of organization mind. In S. Bacharach & N. DiTomasco (Eds.), *Research in the sociology of organizations* (pp. 135-161). Greenwich, CT: JAI Press.

Savage-Rumbaugh, E. S., & Lewin, R. (1994). *Kanzi: The ape at the brink of the human mind*. New York: Wiley.

Schlagger, B. L., & O'Leary, D. D. M. (1991). Potential of visual cortex to develop an array of functional units unique to somatosensory cortex. *Science, 252*, 1556-1560.

Schmandt-Besserat, D. (1992). *Before writing*. Austin: University of Texas Press.

Schmandt-Besserat, D. (1996). *How writing came about*. Austin: University of Texas Press.

Searle, J. (1980). Minds, brains, and programs. *Behavioral and Brain Sciences, 3*, 417-457.

Searle, J. (1984). *Minds, brains, and science*. Cambridge, MA: Harvard University Press.

Searle, J. (1992). *The rediscovery of the mind*. Cambridge, MA: MIT Press.

Segal, G. (1989). The return of the individual. *Mind, 98*, 39-57.

Segal, G. (1991). Defence of a reasonable individualism. *Mind, 100*, 485-494.

Segal, G. (2000). *A slim book about narrow content*. Cambridge, MA: MIT Press.

Segerdahl, P., Fields, W., & Savage-Rumbaugh, E. S. (2005). *Kanzi's primal language: The cultural initiation of primates into language*. Houndsmills, UK: Palgrave Macmillan.

Shapiro, L. (1997). A clearer vision. *Philosophy of Science, 64*, 131-153.

Shapiro, L. (2004). *The mind incarnate*. Cambridge, MA: MIT Press.

Shoemaker, S. (1981). Some varieties of functionalism. *Philosophical Topics, 12*, 93-120.

Simon, H. A. (1981). *The sciences of the artificial* (2nd ed.). Cambridge, MA: MIT Press.

Smith, B. (1999). Situatedness/Embeddedness. In R. A. Wilson & F. C. Keil (Eds.), *The MIT Encyclopedia of the Cognitive Sciences* (pp. 769-771). Cambridge, MA: MIT Press.

Smith, L., & Gasser, M. (2005). The development of embodied cognition: Six lessons from babies. *Artificial Life, 11*, 13-30.

Sober, E. (1993). *Philosophy of biology*. Boulder, CO: Westview Press.

Spivey, M. (2007). *The continuity of mind*. New York: Oxford University Press.

Sporns, O., Chialvo, D. R., Kaiser, M., & Hilgetag, C. C. (2004). Organization, development and function of complex brain networks. *Trends in Cognitive Sciences, 8*, 418-425.

Sprevak, M. (2009). Extended cognition and functionalism. *Journal of Philosophy, 106*, 503-527.

Steels, L. (1994). The artificial life roots of artificial intelligence. *Artificial Life, 1*, 75-110.

Stenning, K. (2002). *Seeing reason: Image and language in learning to think.* Oxford: Oxford University Press.

Stenning, K., & Lambalgen, M. v. (2008). *Human reasoning and cognitive science*. Cambridge, MA: MIT Press.

Stenning, K., & Oaksford, M. (1993). Rationality: Psychological and philosophical perspectives. In K. I. Manktelow & D. E. Over (Eds.), *International library of psychology* (pp. 136-177). London: Routledge.

Stenning, K., & Oberlander, J. (1995). A cognitive theory of graphical and linguistic reasoning: Logic and implementation. *Cognitive Science, 19*, 97-140.

Stenning, K., & Yule, P. (1997). Image and language in human reasoning: A syllogistic illustration. *Cognitive Psychology, 34*, 109-159.

Stephan, A. (2006). The dual role of "emergence" in the philosophy of mind and in cognitive science. *Synthese, 151*, 485-498.

Sterelny, K. (2004a). *Thought in a hostile world: The evolution of human cognition*. Malden, MA: Blackwell.

Sterelny, K. (2004b). Externalism, epistemic artefacts, and the extended mind. In R. Schantz (Ed.), *The externalist challenge. New studies on cognition and intentionality* (pp. 239-254). Berlin & New York: de Gruyter.

Stich, S. P. (1978). Beliefs and subdoxastic states. *Philosophy of Science, 45*, 499-518.

Stich, S. P. (1983). *From folk psychology to cognitive science: The case against belief.* Cambridge, MA: MIT Press.

Strawson, G. (1994). *Mental reality*. Cambridge, MA: MIT Press.

Strogatz, S. H. (1994). *Nonlinear dynamics and chaos: With applications to physics, biology, chemistry, and engineering*. Reading, MA: Addison-Wesley.

Stylianou, D. A. (2002). On the interaction of visualization and analysis: The negotiation of a visual representation in expert problem solving. *Journal of Mathematical Behavior, 21*, 303-317.

Suchman, L. A. (1987). *Plans and situated actions: The problem of human-machine communication*. New York: Cambridge University Press.

Surendran, D. (1996). Measuring the height of trees. *Zimaths, 1*(1). Retrieved from http://www.geocities.com/capecanaveral/lab/3550/tree.htm

Sutton, J. (2006). Distributed cognition: Domains and dimensions. *Pragmatics and Cognition, 14*, 235-247.

Sutton, J. (2010). Exograms and interdisciplinarity: History, the extended mind, and the civilizing process. In R. Menary (Ed.), *The extended mind* (pp. 189-225). Cambridge, MA: MIT Press.

Talmy, L. (2000). *Toward a cognitive semantics: Conceptual structuring systems* (Vol. 1). Cambridge, MA: MIT Press.

Thagard, P. (2005). *Mind: Introduction to Cognitive Science* (2nd ed.). Cambridge, MA: MIT Press.
Theiner, G. (2007). Where syllogistic reasoning happens: An argument for the extended mind hypothesis. In D. S. McNamara & J. G. Trafton (Eds.), *Proceedings of the 29th Annual Cognitive Science Society* (pp. 1551-1556). Austin, TX: Cognitive Science Society.
Theiner, G., Allen, C., & Goldstone, R. (2010). Recognizing Group Cognition. *Cognitive Systems Research, 11*, 378-395.
Thelen, E., Schöner, G., C., S., & Smith, L. B. (2001). The dynamics of embodiment: A field theory of infant perseverative reaching. *Behavioral and Brain Sciences, 24*, 1-86.
Thelen, E., & Smith, L. B. (1994). *A dynamic systems approach to the development of cognition and action*. Cambridge, MA: MIT Press.
Thompson, E. (2007). *Mind in life: Biology, phenomenology, and the sciences of mind*. Cambridge, MA: Harvard University Press.
Thompson, R. K. R., Oden, D. L., & Boyson, S. T. (1997). Language-naïve chimpanzees (pan troglodytes) judge relations between relations in a conceptual matching-to-sample task. *Journal of Experimental Psychology: Animal Behavior Processes, 23*, 31-43.
Tononi, G. (2004). An information integration theory of consciousness. *BMC Neuroscience, 5*. doi: 10.1186/1471-2202-5-42
Tononi, G., & Sporns, O. (2003). Measuring information integration. *BMC Neuroscience, 4*. doi: 10.1186/1471-2202-4-31
Tribble, E. B. (2005). Distributing cognition in the globe. *Shakespeare Quarterly, 56*, 135-155.
Tsakiris, M., S., B., & Gallagher, S. (2007). On agency and body-ownership: Phenomenological and neuroscientific reflections. *Consciousness and Cognition, 16*, 645-660.
Turing, A. M. (1950). Computing machinery and intelligence. *Mind, 59*, 433-460.
Turner, J. S. (2000). *The extended organism: The physiology of animal-built structures*. Cambridge, MA: Harvard University Press.
Turner, V. W. (1969). *The ritual process: Structure and anti-structure*. New York: PAJ Publications.
Unger, P. (1979). I do not exist. In G. F. Macdonald (Ed.), *Perception and identity* (pp. 235-251). Ithaca, NY: Cornell University Press.
Van Gelder, T. (1995). What might cognition be, if not computation? *Journal of Philosophy, 92*, 345-381.
Van Gelder, T. (1998). The dynamical hypothesis in cognitive science. *Behavioral and Brain Sciences, 21*, 1-14.
Van Gelder, T., & Port, R. F. (1995). It's about time: An overview of the dynamical approach to cognition. In R. F. Port & T. Van Gelder (Eds.), *Mind as*

motion: Explorations in the dynamics of cognition (pp. 1-43). Cambridge, MA: MIT Press.
Van Gulick, R. (2001). Reduction, emergence and other recent options on the mind/body problem: A philosophic overview. *Journal of Consciousness Studies, 8*, 1-34.
Van Leeuwen, C., Verstijnen, I. M., & Hekkert, P. (1999). Common unconscious dynamics underly uncommon conscious effect: A case study in the iterative nature of perception and creation. In J. S. Jordan (Ed.), *Modeling consciousness across the disciplines* (pp. 179-218). Lanham, MD: University Press of America.
Varela, F. J., Thompson, E., & Rosch, E. (1991). *The embodied mind: Cognitive science and human experience*. Cambridge, MA: MIT Press.
Vera, A., & Simon, H. A. (1993). Situated action: A symbolic interpretation. *Cognitive Science, 17*, 7-48.
Vygotsky, L. S. (1986). *Thought and language* (A. Kozulin, Trans.). Cambridge, MA: MIT Press.
Vygotsky, L. S., & Cole, M. (1978). *Mind in society: The development of higher psychological processes*. Cambridge, MA: Harvard University Press.
Walsh, J. P., & Ungson, G. R. (1991). Organizational memory. *Academy of Management Review, 16*, 57-91.
Webb, B. (1994). Robotic cxperiments in cricket phonotaxis. In D. Cliff, P. Husbands, J. A. Meyer & S. Wilson (Eds.), *From animals to animats 3: Proceedings of the Third International Conference on Simulation of Adaptive Behavior* (pp. 45-54). Cambridge, MA: MIT Press.
Weiser, M. (1991). The computer for the 21st century. *Scientific American: Special Issue on Communications, Computers, and Networks*, 94-104.
Weiskopf, D. (2008). Patrolling the mind's boundaries. *Erkenntnis, 68*, 265-276.
Wertsch, J. V. (1998). Mediated action. In W. Bechtel, G. Graham & D. A. Balota (Eds.), *A companion to cognitive science* (pp. 518-525). Malden: Blackwell.
Wheeler, M. (2005). *Reconstructing the cognitive world: The next step*. Cambridge, MA: MIT Press.
Wheeler, M. (2010a). In defense of extended functionalism. In R. Menary (Ed.), *The extended mind* (pp. 245-270). Cambridge, MA: MIT Press.
Wheeler, M. (2010b). Minds, things and materiality. In C. Renfrew & L. Malafouris (Eds.), *The cognitive life of things: Recasting the boundaries of the mind* (pp. 29-37). Cambridge, UK: McDonald Institute for Archaeological Research Publications.
Wheeler, M., & Clark, A. (1999). Genic representation: Reconciling content and causal complexity. *British Journal for the Philosophy of Science, 50*, 103-135.

Wilson, M. (2002). Six views of embodied cognition. *Psychonomic Bulletin and Review, 9*, 625-636.

Wilson, R. (1994). Wide computationalism. *Mind, 103*, 351-372.

Wilson, R. (1995). *Cartesian psychology and physical minds: Individualism and the sciences of the mind*. Cambridge, UK: Cambridge University Press.

Wilson, R. (2001). Two views of realization. *Philosophical Studies, 104*, 1-31.

Wilson, R. (2004). *Boundaries of the mind*. New York: Cambridge University Press.

Wilson, R. (2005). *Genes and the agents of life*. New York: Cambridge University Press.

Wilson, R. (2006). *Commentary on Robert Rupert, "Extended Cognition As a Framework for Cognitive Science: The Costs Outweighs the Benefits"*. Paper presented at the Central Division Meeting of the American Philosophical Association, Chicago, IL.

Wilson, R. (2010a). Meaning making and the mind of the externalist. In R. Menary (Ed.), *The extended mind* (pp. 167-188). Cambridge, MA: MIT Press.

Wilson, R. (2010b). Extended vision. In N. Gangopadhyay, M. Madary & F. Spicer (Eds.), *Perception, action and consciousness* (pp. 277-290). Oxford: Oxford University Press.

Wilson, R. (2010c). Review of Robert Rupert: "Cognitive systems and the extended mind". *Notre Dame Philosophical Reviews, 2010.03.07*. Retrieved from http://ndpr.nd.edu/review.cfm?id=19128

Wilson, R., & Clark, A. (2009). How to situate cognition: Letting nature take its course. In M. Aydede & P. Robbins (Eds.), *The Cambridge Handbook of Situated Cognition* (pp. 55-77). New York: Cambridge University Press.

Wimsatt, W. C. (1986). Forms of aggregativity. In M. G. Grene, A. Donagan, A. N. Perovich & M. V. Wedin (Eds.), *Human nature and natural knowledge: Essays presented to Marjorie Grene on the occasion of her seventy-fifth birthday* (pp. 259-291). Dordrecht: Reidel.

Wood, D., Bruner, J. S., & Ross, G. (1976). The role of tutoring and problem solving. *Journal of Child Psychology and Psychiatry, 17*, 89-100.

Yokoi, H., Hernandez, A., Katoh, R., Yu, W., Watanabe, I., & Maruishi, M. (2004). Mutual adaptation in a prosthetics application. In F. Iida, R. Pfeifer, L. Steels & Y. Kuniyoshi (Eds.), *Embodied artificial intelligence* (pp. 146-159). Berlin: Springer.

Zednik, C. A. (2008). Dynamical models and mechanistic explanations. In B. C. Love, K. McRae & V. M. Sloutsky (Eds.), *Proceedings of the 30th Annual Cognitive Science Society* (pp. 1454-1459). Austin, TX: Cognitive Science Society.

Zhang, J., & Norman, D. A. (1994). Representations in distributed cognitive tasks. *Cognitive Science, 18*, 87-122.

Zhu, L., & Gigerenzer, G. (2006). Children can solve Bayesian problems: The role of representation in mental computation. *Cognition, 98*, 287-308.